"[One of] America's nine funniest writers. . . . A spot-on favorite with women who grew up during the 1960s and 1970s, Kaplan at her best reads like a black-market episode of *The Wonder Years*." —*Southwest Airlines Spirit* magazine

"*Why I'm Like This* is an impressive debut, and Kaplan has a wonderfully natural comic style." —*Detroit Free Press*

"Kaplan goes straight to the funny heart of things. . . . I can't imagine a mother who wouldn't die laughing." —*New Orleans Times-Picayune*

"Kaplan is smart, funny, brave, and totally original. It's an amazing collection, one you'll want to read again and pass on to all of your friends. You'll think you're reading David Sedaris; it's that good." —BookSense.com

"Nightmarish and hilarious. . . . Written with such fierce denial and tenderness. . . . [*Why I'm Like This*] made me cry out loud." —Chicago Public Radio's *Network Chicago*

"Kaplan makes it perfectly clear why she's like this—with an acerbic grace all her own. . . . Even if where she's taking you is someplace you've yet to arrive or will never go in her capacity, you will feel as if you are there alongside her. . . . She's that good." —*News Enterprise* (Hardin County, Kentucky)

"A fine achievement, *Why I'm Like This* will reach a wide and appreciative audience." —*Midwest Book Review*

"Cynthia Kaplan is so delightful a writer, so funny and smart and tart, I'm sure her family won't even notice that she's turned on them." —Dorothy Gallagher, author of *How I Came into My Inheritance*

"With her unique brand of humane observation and wit, Kaplan has written a book for the ages, a book that will no doubt be passed on, from hand to hand, in family after family, by people who recognize some heretofore-unearthed part of their own history or selves in the stories she tells."

—Beth Kephart, author of *A Slant of Sun*

"Funny, sweet, weirdly life affirming, and painfully true. . . . Cynthia Kaplan's prose generates the rarest kind of laughter—the kind that makes you cringe in recognition, then thank the writing gods that somebody else out there gets it."

—Jerry Stahl, author of
Permanent Midnight and *Plainclothes Naked*

"If you have a best friend who is the funniest person you know, who you call just to cheer yourself up, her name must be Cynthia Kaplan. The rest of us can happily make due with reading this delightful book."   —Cathleen Schine, author of *The Love Letter*

"Cynthia Kaplan's gift is her ability to take on life's absurdities and come out a winner. Her stories are sharp, touching, and deliciously ironic, but most of all they are true in every sense of the word. *Why I'm Like This* is a rollicking tour through the wild and crazy landscape of today's world—a necessary book for all of us who sometimes question why we're like this."

—Debra Ginsberg, author of *Waiting* and *Raising Blaze*

"A self-assured, unified work that's sexy in the best sense: mature, candid, and real. Often compared to David Sedaris, [Kaplan], the actress/writer, combines droll humor with hard-won sentiment."

—*Seattle Weekly*

"What makes [Kaplan's] true stories so compulsively readable are [her] razor-sharp prose and contagious sense of humor. . . . Kaplan's writing is laugh-out-loud funny. Right up there with nonfiction writers such as David Sedaris and Cynthia Heimel."

—Richmond.com

why I'm like this

TRUE STORIES

CYNTHIA KAPLAN

HARPER PERENNIAL

NEW YORK • LONDON • TORONTO • SYDNEY

AUTHOR'S NOTE:

I have changed the names of some people and
some places because my editor made me.

HARPER ● PERENNIAL

"Better Safer Warmer" was originally published in the *New York Times* on February 1, 1999.

A hardcover edition of this book was published in 2002 by William Morrow, an imprint of
HarperCollins Publishers.

P.S.™ is a trademark of HarperCollins Publishers.

HarperCollins books may be purchased for educational, business, or sales promotional use.
For information please write: Special Markets Department, HarperCollins Publishers,
10 East 53rd Street, New York, NY 10022.

First Perennial edition published 2003.
First Harper Perennial edition published 2007.

Designed by Bernard Klein

The Library of Congress has catalogued the hardcover edition as follows:
Kaplan, Cynthia.
Why I'm like this / by Cynthia Kaplan.—1st ed.
    p. cm.
ISBN 0-688-17850-2
    1. Kaplan, Cynthia—Anecdotes. 2. Entertainers—United States—Anecdotes. I. Title.
PN2287.K23 A3 2002
791'.092—dc21   2001058657

ISBN: 978-0-06-128396-3 (pbk.)
ISBN-10: 0-06-128396-7 (pbk.)

07 08 09 10 11 ❖/RRD 10 9 8 7 6 5 4 3 2 1

For David

contents

———

The eyes are the windows of the head.

queechy girls

—

THERE was always one girl at camp whom everyone hated. It had nothing to do with cliques or teams or personal dislikes, and it was not even that everyone had discussed it and a consensus had been raised based upon certain irrefutable evidence. It was just like everyone hated lima beans and the color brown. It was obvious and it was universal, so it didn't require organization.

Everyone at Queechy Lake Camp hated Lisa Hope Mermen. There were no reasons why and there were a million reasons why. Her breasts were too large and her hair was limp. She had probably had her period since she was ten. She was a *very* mediocre athlete. She was not nor ever would be considered coltish. She was nice to everyone and some people hate that. She had no friends and some people took that as a sign. She had two first names and insisted on using both. At best, she was ignored. At worst, she was teased and bullied and shoved into the lake. Tricks were played on her, food stolen from her. Intimate articles of her clothing, particularly her brassiere and large to-the-waist panties, were raised on the flagpole in the morning just before assembly. There they were buffeted unkindly by the Maine breeze, these colors of the enemy territory, to be saluted by smirking, suntanned cuties.

Why was she still here, Lisa Hope Mermen? Why did she return summer after summer to a camp where a philosophy of equality symbolized by a de rigueur camp uniform of simple white midi blouses and navy shorts still failed to work in her favor because *her* midi blouse required darts? Why didn't her parents switch her to music camp or send her to Europe where everyone had limp hair?

Queechy Lake Camp was certainly the most beautiful girls' camp in Maine. It was situated on a tree-topped hill which gracefully sloped down to the edge of the lake, clear, blue-black, and serene. At the high end of the camp the bunks formed a large circle around a perfectly manicured blanket of grass, unlike the bunks at Pine Forrest and Bluebird Lake, which were dotted willy-nilly throughout the woods. At the center of the circle was the aforementioned flagpole. As night fell, this happy configuration of lodgings, their lights

winking in the dusk, resembled nothing so much as a shoreline of exclusive summer cottages; the darkening courtyard,
a navy lake. Paradise.

Then, a little lower down, there was Queechy House, an
enormous green Adirondack affair standing exactly as it had
for almost one hundred years, its plaque-covered walls a tes-
tament to the overachievements of past Queechy girls: Best
Field Hockey, Best Basketball, Best Waterskiing. The Archery
Award. The Craft Award. The Queechy Spirit Prize. On one
side of Queechy House there were the living room and the
commissary and the mail room, and on the other side were
the dining room and the kitchen (which no one ever saw
except on Cinnamon Toast Nights, when everyone in your
bunk got to go in and eat as much cinnamon toast as they
could. The record, set by Rose Bunnswanger in 1957, was
something like forty-nine pieces). Behind Queechy House
was a gathering of humongous old pine trees and beneath the
trees were twenty or so Adirondack chairs, painted dark
green. This spot was called Beneath the Pines, without sar-
casm. Team rallies happened here, and Friday night services.
If you were friends, one of you sat in the seat and one of you
perched on the wide armrest, so you were connected, so
there was no mistaking it.

Every building had a name. Please Come Inn and Nellie's
Nest and The Barn and Hill House and Mildred, *just* Mildred, after an English lacrosse counselor who perished in the
bombing of Dresden. She had gone there with false papers to

search for two elderly cousins who were believed to have been in hiding. There was a field hockey field and a lacrosse field and two softball fields and there were tennis courts and volleyball courts and basketball courts and sailing and canoeing and waterskiing. And there was kickball and newcomb for the younger girls. The field hockey and lacrosse counselors came from England, like Mildred, because the English know those sports best.

There were no socials or dances with boys' camps because Queechy girls were renowned for their winning combination of athletic ability, teamwork, and pep, and pitting them against each other for the attentions of pimply-faced, perpetually engorged (that's Deb Edelstein's word, not mine) boys from, say, Camp Tonkahanni, might undermine the confidence of even the most spirited, talented Queechy girl, not to mention threaten many deep friendships. I, for one, was perfectly happy not to have to deal with some dopey tennis nerd trying to guess my bra size. There were male counselors, of course, but it was not the same because they were all over eighteen. There was Bill Ski and Mark Ski and Jamie Canoe and Jack Tennis and Bill Tennis and Chris Swim and Mike Softball. There was Somebody Riding whose name I never remembered because I hated riding. And there was Corey Silver Shop whom everyone assumed was gay even though most of us had no idea what we were talking about.

There was a theater called Marion's Tent, though no one remembered Marion and there was no tent. My theatrical career at Queechy Lake Camp was distinguished by many memorable performances as the second lead; a girl named Wanda Massey always got the starring role. For seven summers, I was the male half of nearly every romantic coupling written for the musical theater. I was Oscar to her Charity, Captain Von Trapp to her Maria, Tony to her Maria. She was, metaphorically speaking, *always* Maria.

The very moment parts were posted and scripts handed out, each of us would rush to the edge of some grassy slope to count our lines. The more lines, the bigger the part. Plot development and character were irrelevant. This summer we were doing *The Miracle Worker.* Wanda Massey was Annie Sullivan and I was Helen Keller. When I opened the script and saw that I had only one line for the entire first half of the play, and that line consisted of one word, "wawa," I nearly went berserk.

Once a summer the Story of Queechy Lake Camp was retold by Aunt Jeanne, the camp director. The entire camp gathered Beneath the Pines, a tangle of interlocked arms and legs, with much tickling of forearms and backs and braiding of hair, to hear how during World War I the camp planted its playing fields with navy beans. Girls as young as nine years old pushed the seeds into the soil and two months later plucked the beans from the leafy vines. They sewed shirts and knitted socks. They were industrious and patriotic and occasionally had air-raid rehearsals. They wore baggy bloomers and smocks and maybe in those days it didn't matter if you were bad at sports or had a large bosom. Some of these girls were the grandmothers and mothers and aunts of future Queechy girls. Katie Cohen was a legacy and so was Beth Reingold and so was Deb Edelstein. Tessie Green's grandmother won the spirit prize twice in a row and she was completely deaf in one ear. Joan Grobman's mother lost half her middle finger in a rock-climbing accident in 1968, the summer she was fifteen, and came back to camp as soon as she was out of the hospital. That's what Queechy girls did. When the story was over everyone sang Queechy Lake songs: "Spirit of Queechy," "Far Above Dear Queechy Waters," "Queechy Friends Forever."

Surely Lisa Hope Mermen was not the only girl whose bathing suit, with its built-in brassiere, remained dry on the front following the backstroke race at swim meets. Surely she was not the only girl without a lilting voice or curly ringlets. And surely a lilting voice and curly ringlets were not the only prerequisites of a successful adolescence. Although who was I to say, since I had a reasonably lilting voice and a decent head of curly ringlets?

But did Lisa Hope Mermen really look all that unhappy? She cheered from the benches at softball and field hockey games, gave her ineffectual all on the "B" basketball squad, sang her heart out in the chorus of *Call Me Madam*. She befriended younger girls who either didn't know any better or were similarly ill suited to the demands of popularity.

Counselors were sympathetic. I was bewildered almost out of my complacency. Almost.

I'd had too many s'mores. I had a weakness for them. I liked the marshmallow to catch fire and burn the entire outside black. I liked the Hershey's chocolate to still be hard, even a little cold; it had to hold its own dually against the heat of the marshmallow and the firm crunch of the graham cracker. If you ate a s'more during a lunchtime campfire, chances were the chocolate bar would be melty from sitting in the sun. Then it melted even more when it came into contact with the marshmallow, and suddenly the graham cracker dominated. I hated that. I'd just as soon skip it if that was how it was going to be. At night, though, on the beach, when everything cooled down, the grains of sand a silver trickle between your toes, when the lake met the sky, when hooded sweatshirts were in order, *that* was the time to gorge yourself into oblivion on s'mores.

I'd had seven, which as anyone knows is four too many. And actually I was feeling all right until the camper–counselor game of duck duck goose. A few woozy moments after I'd chased Karen Basketball two loops around the circle, I felt the s'mores rising up. I felt a revolution of s'mores. I ran up the beach and gave them their freedom behind the sailing hut. I looked down to inspect the damage (everyone looks at their vomit, everyone) and catch my breath. Running and then vomiting is harder than just vomiting. I leaned against the side of the hut. My breath was loud in my ear. Too loud. And there was a funny humming noise. *Mmmmm. Mmmmm.* It took me a minute to realize it wasn't coming from me. I inched forward, careful not to step in my pile of ex-s'mores, and moved around the corner of the sailing hut until I was crouching beneath its one open window, just next to the door.

At first I thought it was, I don't even know, just this big thing, this moving shadow, it was so dark in there. *Oh, God, it's a bear. Shit, fuck, shit.* But then, *then*—in that way you can see people in the movies when there are scenes in the dark, like there's an illogical light source but you accept it because it's the movies and you really want to see what's going on—*then*, I saw her. Years later I realized the moon must have broken through some clouds.

Lisa Hope Mermen lay all but naked on a pile of sails. Her brassiere was wrapped like a bandage around one arm, her dark blue camp shorts scrunched between her legs. Her white body, all breasts and belly and thighs, was aglow. She was breathing heavily; her face looked . . . I didn't know. Like she was constipated but happy about it. What was she doing? My legs, already vomit-wobbly, were starting to ache from crouching at the window, but I couldn't tear my eyes away. I was watching the Lisa Hope Mermen Movie. The humming got louder. Why was she lying there, naked, humming? *Why?* It wasn't even a song. Then, suddenly, the dark shorts between her legs were moving and I saw that they weren't her shorts. It was like realizing a piece of mud stuck to your ankle is really a leech. The thing between Lisa Hope Mermen's legs rose up and smiled. A glistening, mustached smile.

Mark Ski.

He moved out of the shadows, naked, his back to me, and stretched his neck to both sides. She said, "That was yummy." He said, "Good." There was a rushing noise in my ears. He lowered his whole body onto Lisa Hope Mermen very, very slowly. I saw his penis. My legs collapsed beneath me. I dropped onto the sand. Poomf.

When I returned to the beach everyone was sitting Indian style in a big circle around the fire and singing "Leaving on a Jet Plane." Joan Grobman and Deena Saks made a place for me between them. I guess I sang along.

A substantive discussion of What I Saw did not commence until after the campfire died and everyone more or less headed off to bed. It lasted until well after two and included my own eyewitness testimony, followed by a question-and-answer period and concluding with a sort of fake, sort of real hypnosis session, in case there was something I was repressing. Then we tried to levitate Beth Reingold.

We slept in pairs in canvas tents on the sand. Lying in my sleeping bag beside Deb Edelstein, her soft asthmatic wheezing keeping time in the dark, I realized this would be the last time. In the morning we would have blueberry pancakes and hot chocolate and go skinny-dipping in the glassy, dawn-cold lake, and then head off up the hill to our various scheduled activities—lacrosse or archery or pottery. But this was the last senior overnight. The last campfire on the beach. And there was only one more swim meet to go and one more day trip to Acadia National Park and maybe one more pajama breakfast, if we were lucky. There was only the counselor show, *Man of La Mancha,* left to see. All the craft projects would have to be finished in the next week, all the bunk food eaten, all the lost things found. Suddenly it would be the very last night, the night of the Senior Serenade, when we would go by flashlight from bunk to bunk, like carolers, singing the old songs. I was not going to come back the next summer as a junior counselor because my parents wanted me to do a summer session at Andover. And anyway it wouldn't have been the same. When something's over, it's over.

a dog loves a bone

———

HAD I known what high school would be like I would have asked my parents to set me adrift on an ice floe at puberty. You can be scarred for life by the fact that no boy in a leather-and-wool varsity jacket held your hand in the twilit ring around the football field at a night game. Or combed your long hair out with his fingers. Or tackled you repeatedly during snowball fights. No level of personal or financial success will ever alter the fact that the tousle-haired, sly-smiling boy who sat two rows behind you in trig never played catch with your flute with his friend, and then, when he gave it back to you, draped himself on your locker door and said, "So, why don't you play me something?" And even if, by some completely anomalous blip in the world order, the tousle-haired, sly-smiling boy actually *did* steal your flute or, if you didn't play the flute, your biology book or your hat with the pom-pom, and then asked you the question that wasn't really a question but something else, some spark just before the moment of ignition, but you didn't get the gist of the whole thing, had no experience, were too taken off guard to get it right, if you said, "Oh, uh, well, I don't know, um, sure, I guess . . . ?" and then he just smiled his sly smile and said, "Forget it, see ya in trig," all this being actually worse than no chance, this being the rare chance blown, as I said, you can be scarred for life.

When I was a junior in high school, Kevin Parker asked me to slow-dance to "House on Pooh Corner," a song I knew all the words and guitar chords to. My normal range of activity at high-school dances consisted mostly of standing around with my friend Mary Beth, making fun of the cheer-leaders French-kissing the jocks during "Stairway to Heaven." "They're not even dancing," we would say. Occasionally I would fast-dance with a boy on the fringe of cool or on the JV basketball team, or clump up with my gang of girlfriends and try to look as if we were all having the time of our Honor Society lives. I had adored Kevin Parker since the early grades, lapsing only briefly to have a crush on his friend Mike Petersen, a boy to whom I'd spoken maybe ten words but whose smile made my hair hurt. Mike Petersen's family were Mormons and kept a year's worth of canned goods in their basement in anticipation of the Apocalypse. I used to ride my bike past their house on River Road and stare at it in wonder. It was so small; a year's worth, wow. At the end of sophomore year they moved to Salt Lake City and my affections reverted back to their original object.

So, one night at a school dance, Kevin Parker, between girlfriends, or maybe just out of pity, maybe it was a pity dance, took me by the hand and like Moses leading the Israelites from bondage, led me from the shadows of the folded wood bleachers out onto the light-dappled (mirror ball) gym floor.

He put his hard body right up to mine and dropped his blond head near my shoulder and ran his hands all over my back and up and down my sides. What was I supposed to do? No, really, what *was* I supposed to do? My head was on his chest and my arms around his neck and I was a frozen-solid block of ice. Didn't move a muscle above my waist the entire song. I was like a robot who has come to the end of its capabilities. I frantically probed my engines for further instructions. There was only this: slow-dancing comes with a different model.

As the last strains of "House on Pooh Corner" faded into the opening chords of "Brown Sugar," Kevin Parker whispered, not unkindly, in my ear, "You can wake up now." I was fifteen.

I'm going to make a suggestion. You can take it or leave it but here it is: try never to give a man the impression you are asleep when you are not. This may change one day, but for now it is sound advice.

That summer, I taught canoeing at a YMCA camp on the Aspetuck River. For weeks before camp began my mother repeated over and over that the son of an old college friend would be working there, too, and that I should look out for him. I could vaguely recall the son of another of her college friends as being short and spindly and beady eyed, and I was so tired of hearing his name, Jamie Karlin, Jamie Karlin, that I vowed to have nothing to do with him on principle, the principle being to reject anything my mother ever suggested. Little did I know that soon I, too, would be repeating, no, chanting his name over and over and over and over again, until the vowels and consonants had bored themselves into my skull as into a tree trunk, the deep grooves of which I might trace for the rest of my natural life.

Jamie Karlin was heartbreakingly beautiful. Looking at him was like looking into the sun. I could only do it for a moment or two at a time before my eyeballs burned and my head ached. He was a golden shining thing, a supernova, a young lion. To break him down into his parts, his wavy blond hair, the golden, fleecy cilia that covered his arms and legs, his *mouth*, would be counterintuitive; you wouldn't parse a rose or a bald eagle or the last day of school.

He taught woodworking. One day he almost cut off his finger on the lathe and when he came home from the hospital that night, tired and sweet on painkillers, the other boating counselor, Chris, and I watched TV with him at his house. At one point for no reason he reached over and stroked the back of my calf. Other people have memories of making out in rec rooms. What can I do?

Then, on a balmy night in late July, Jamie Karlin took me to a party, and afterward, while sitting in my driveway in his car, I nervously popped the door handle, engaging the car's interior light. "Isn't that kind of bright?" he whispered. "Sorry," I whispered back, and then I said, "Good night" and "See you tomorrow," and I got out of the car. No one ever told me that when a guy says it is too bright he means I want to kiss you, not Get out of the car and shut the door behind you.

And yet, the summer was not a complete disaster. On one of the last true blue days in August, when the oppressive midsummer haze had been blown away by some zippy cumulus clouds, opportunity knocked once more. At the counselors' pool party, Jamie Karlin asked me to rub sunblock onto his back. He had the most unbelievable, fair, freckly skin. Had I this moment to live again, I would have offered to apply the sunblock later, in private, with my tongue. At the time, though, I was too overcome to even speak. It was all I could do not to pass out. Mute, I sat down next to him in the cool grass that sloped above the pool and caressed his lanky, soccer-boy body with Coppertone.

I was obsessed with Jamie Karlin for years after. *Years.* I loved him like a dog loves a bone. Why do they do that? There is no meat left on it. Is it wishful thinking? Is it the idea that meat was there once and maybe it will be there again one day? Or is it just nostalgia? Oh, that meat was good, remember that meat? Nummy nummy nummy. There I was, chewing my love down to a nub and then burying it and then digging it up and then burying it again somewhere else. And then digging it up. Didn't want any other dogs to find it! If nothing else, I was loyal.

And there *was* nothing else.

I did not have a romantic relationship with any of the great loves of my youth. In college, when boys I liked started liking me back, that is, when the moon was in the seventh house and Jupiter aligned with Mars, I often didn't know what to do with them. Still, I made out with a reasonable number of athletic, slightly grungy, flannel-shirted guys, because they were the kind I'd grown up with, the kind I couldn't have back then, but since my university was eight trillion times larger than my high school, the numbers were in my favor. Unfortunately, none of these boys turned out to be boyfriend material. Somehow, along with lanky and sexy, it seems I was often attracted to dumb. Well, not dumb, but *unwitty*, which is not a prerequisite for firsts but makes it mighty hard to go back for seconds. *Therefore*, I fell hopelessly in love with a series of smart, witty fellows who either weren't interested in my virginal self or scared me because I was my virginal self. Perfect. Nummy.

To top it all off, I was always friends with the girl all the men wanted. I'd like to say it was by accident, although that sounds naive, but I was naive. It certainly didn't do me any good. Perhaps these girls sought *me* out. I was classic side-kick material, cute enough to be seen with but not cuter than you. My best friend at Andover summer school was a tall Texas girl with blond hair and stick-straight legs. She looked amazing in tennis shorts and was adored by all. I was her first Jewish friend. She'd never even eaten a bagel. We were like Candy Bergen and, I don't know, some short Jewish girl.

My first college roommate, randomly assigned to me, turned out to be a siren, beckoning men to their destruction with her arching eyebrows and C-cups. When her boyfriend from home visited her they had noisy almost-sex half the night while I had to pretend to sleep. She made out in the stairwell of our dorm with a guy I liked. She had walked out with him with the intention of telling him how great I was and then, she said, he jumped her! Imagine that!? Drunk, she went home from a party with another guy I liked, and after they dry-humped he threw up on her. Sure, I had to laugh, it was funny. Ha ha ha. We were inseparable.

I finally had sex and it didn't change anything. Or rather, it changed everything. As good as sex was, it had a way of bursting the bubble. My early loves were based on a stom-ach-roiling mixture of visual cues and wishful thinking. I wanted something that was substantial only in my imagining of it. Domesticity, even the barest minimum that is a result of waking up together and requiring showers and food, felt awkward and contrived. This is because after you start having sex, you go through a period where it is easier to have sex than it is to have a conversation, which could and ultimately did lead to the unpleasant discovery that the object of your attraction is an idiot. Wait, we've come full circle; sex didn't change anything again.

There was a guy I slept with for five months. We had almost nothing to talk about but he could come to orgasm twice without ever pulling out, like that Doors lyric, "Love me two times . . . I'm goin' away." I was both sad and relieved when he ended it. At the time, the acid test for me was whether or not a guy had read *Ethan Frome,* and he failed. One night he fell asleep on the bench in his entryway with a piece of toast in his hand.

I had a boyfriend who yelled at me for mopping his bathroom floor because I couldn't stand to walk barefoot in there. He wrote a terrible play about boarding-school jocks sodomizing the sensitive freshman and didn't give me a part in it. He thought art for money was crass so he lived in an apartment his parents rented for him and eventually ended up in TV. He went to sleep listening to sports radio, like my grandfather listened to the stock quotes. He broke up with me by disappearing for two months without a word.

Then I had a boyfriend who routinely abandoned me at his friends' parties, parties where I knew no one. Unless someone interesting was paying attention to me and then he would slither over me, like a creeping vine, to share the light. He would seduce me and as soon as I was seduced he would complain that he felt boxed in. He forgot to bring his credit card to the restaurant on Valentine's Day and then never paid me back. He took an eternity to come. The day I figured out that sticking my finger in his anus did the trick I bought myself a new hat. I tortured myself with him for a year, on and off. The off was when I intuited that he was sleeping with a colleague at his glamorous, high-powered job. Not a remarkable feat, the intuition, that is, since I heard him talking to a friend about her on the phone, sorry, *my* phone, while he thought I was still in the shower, sly dog. The woman was wafer thin and had long red curlicue hair. Every man I know was in love with her at some point or another. So, of course, we became friends. I gave the guy a second chance but it ended anyway when I came to the realization that his grammatical errors would eventually drive me out of my gourd.

And there were other guys all during the black hole that was my twenties. Like the one who stood me up at my birthday dinner. And the one who wanted to caress my breasts until I thought my head would explode. (Note: Head explode is not a euphemism for something good.) And there was the handsome lawyer from Baltimore. He had a big truck. No, really, a truck.

Why would I have sex with such people? Beats me. I had to have sex with somebody. I needed practice/I got a late start/it felt good. I couldn't spend my whole life waiting for some soul-crushingly beautiful boy to give me just enough rope to hang myself with. At some point I had to let there be more. Sex is what grown-ups do with their pent-up longings. They don't just pine away; they can't be nostalgic for something they've never had. Well, I can, because I have a very good imagination, but most people can't.

I was so used to the agony and anxiety of unrequited love that I thought that agony and anxiety *were* love. It wasn't enough for someone to make my pulse race, he had to make me sick. One boyfriend, Mr. My Filthy Home Is My Castle, actually gave me migraines; he was my hero. The pain of it all was also an antidote to the tedium, masking the most banal of connections with time-sucking, brain-freezing uncertainty. Why talk about politics or books or anything when you can talk about whether or not this is a relationship? Or why not just have sex. Ah, again you see we've come full circle.

My stomach finally stopped hurting the morning after my first date with the man who would become my husband. I woke up at seven kicking my feet under the covers, a little scissor-kick dance, a dance of lightness and ease. I did not know then that I would marry him, but I knew that I would be all right. There was a whole new breed of man out there, a gold standard, and I had finally tapped the vein. It was actually possible to go out to dinner, go home and have sex, wake up, shower and have breakfast, conversing *throughout*, all without being either bored to tears or in tears. Of course, there are still times, yes, when I wish I had to worry whether or not he is going ask me to dance or sit next to me on the bus on the way home from the class trip to Mystic Seaport. There are times when I wish I could worship him from afar and listen to Paul McCartney's *Ram* twelve times a day because it is his favorite album. I miss the idea of the agony.

But that's okay. Because sometimes, at night, we pretend that he is a border guard and I am a graduate student in archeology, and I am able to make do with that.

this is for you

Do you love this? If you don't love it, don't take it. I'll give it to your cousin, because she will love it. How about this? I bought this in Singapore/Madrid/Palm Beach/ Mexico City, in the gift shop of the *Queen Mary*/the Bel Air/the Georges Cinque/the Hong Kong Hilton. It is made of the best jade/ivory/gold and has little diamonds/garnets/ rubies/seed pearls spelling out Lillian/LSS/Dearest. Take this little fish/heart/chai/elephant with the tusk up. It represents luck/life/love/wealth/health. I want you to have this. And this. One day.

When I'm dead.

* * *

Every time I visited my Florida grandmother, my mother's mother, Lillian Siegel, we made a full account of her jewelry. We would sit together on her bed and she would display her wares. Each necklace and ring, each bangle, each pin had a back-story, a provenance. She wore the baroque pearls on the ship the *France* and the captain had a mad crush on her. She haggled with a blind man in a shop in India for the ivory charm and surely there was not another of its kind in the world. She bought the ring with the serpent that climbed up your finger in a hotel in Beverly Hills just before she rode the elevator with Cary Grant.

"Oh, Mr. Grant, how do you do? Do you know that I've seen all your pictures twice?"

My grandmother was an extremely charming woman and I have no doubt that Cary Grant chatted with her all the way from the lobby to the penthouse. Men often attached themselves to Lil, even though she was married. Suave young men of Latin origin wrote her letters, addressing them *To my dearest friend* and signing them *With undying gratitude*. A priest she met in Singapore relied on her for spiritual advice. Dance instructors on cruise ships invariably forsook all other students so that they might dance the merenge each night with her. She was not a flirt, rather she was open-minded, lighthearted, intelligent, amusing. She made friends and had admirers wherever she went.

Lil often traveled without my grandfather, Ben, whom she had met as a teenager in Brooklyn. They required time apart;

they argued every day of their married lives, and they were married three times. Once at seventeen, before it was legal; once at eighteen, in front of their families; and once again in the seventies, after they had been divorced for a year or two. While they were divorced my grandfather had a brief, inter-mezzo marriage to a woman named Ruth who clearly wasn't up to the task. Too docile. My grandparents' divorce was, by my mother's account, an acrimonious affair, and during the proceedings Lil was offered the services of a retired gangster who was living on the other side of South Ocean Boulevard at the Diplomat Towers. She graciously declined his pro-posal to fit my grandfather with a pair of cement shoes and heave him into the Inter-Coastal Waterway, but she did accept a gold-plated pocket watch from him, which, during one of our sessions, she gave to me.

Lil liked everything around her to be special. She thought each and every one of her belongings was exquisite, and she would ask you, when you visited, if this or that wasn't the most exquisite you'd ever seen. As I got older it became clear that it was more a matter of it being the *only* I'd ever seen. Certainly no one else's grandmother had glued little ornate pillboxes to the tops of a pair of Chinese urns to give them "interest." Or crocheted a border onto her bathroom rug. Or twined a garland of silk ivy around the naked body of a lamp nymph. Or sawed a couple of inches off some valuable din-ing-room chairs that seemed too high. Lil had only a cursory respect for a thing's intrinsic value. It's an antique? So what?

An ugly drawer pull is an ugly drawer pull. What it needs is gold leaf. My mother had a little inlaid-wood side table that I coveted. It had one tiny drawer which while I was growing up housed a single old-fashioned skate key, conjuring up fantasies involving a boy from my high-school hockey club. The table was one of a pair belonging to Lil, my mother told me, but she didn't know what had happened to its mate. On one visit to Florida, I noticed that a small side table with a top like a large porcelain platter had the same dark, three-legged base as the inlaid table. I asked my grandmother about it. Oh, she said, she had taken off the wood top and glued in its place a porcelain platter. Wasn't it exquisite?

In my grandparents' kitchen there was a charming glass-topped table and two chairs made of wrought iron and painted pink. On each chair there was a thick fuchsia-colored vinyl cushion and every time you sat down it released a sustained hiss, maybe fifteen seconds long, as it deflated beneath your weight. When you stood up it stuck to the back of your thighs, then slowly peeled away, making a sound like tape being ripped off a package. The price of beauty.

My grandmother became known in her various apartment complexes in Hallandale for her handmade three-quarter-sleeve sweaters, or "bracelet sleeve," which meant the arms were short enough to show off an elegant wrist and whatever exquisite thing dangled from it. Lil and I cataloged these, too, and my mother and I always wore them when we visited, to please her, although on us the sleeves did not look elegant, just too short. There were probably more than forty sweaters, and they fell into several groups. There was the solid-pullover-with-the-little-V-neck-and-gathered-shoulders group, the all-over-tweedy-color-mix-with-contrasting-cuff-cardigan-and-sometimes-a-sparkly-metallic-yarn incorporated group, and there was the variegated-stripe group, which consisted of sweaters with a contrasting stripe on one arm but not the other, or on one panel but not the other. Or stripes in front but not in back, or in back but not in front. Sometimes a whole sweater was red except for one white sleeve with a red cuff. And there was often some extra crocheted detailing. Like a little pocket, or a collar or loopy hem. Actually, my grandmother gave her store-bought jackets and hats and sometimes even her tablecloths the same treatment—a little crocheted border. She even held knitting classes poolside. All over South Florida there are scores of little old ladies in color-block sweaters with the sleeves too short, and just as many relatives up north too guilt ridden to give them away and too embarrassed to wear them.

My grandfather's main interests were his grandchildren, his stocks, and the beach, which he walked every morning. He had a deep tan and a head full of white hair, which I loved to comb, until my father, as a joke, told me it was a toupee. In his day, Ben had been what was called a natty dresser, but his fashion sense in retirement could only have been interpreted as a conscious choice to antagonize his wife. Although they had plenty of money, he took to wearing slightly grungy, often loud sport shirts and mismatched shorts or bathing suits, and white loafery types of shoes, dirty canvas boaters, or worn-down flip-flops. I think he even owned a clip-on tie. No, I'm sure he owned a clip-on tie. I can suddenly see him standing in the apartment, handsome even in light blue leisure-suit-type pants and a short-sleeved oxford, clipping the damned thing on. My grandfather had the most wonderful look, as if he had boxed in his youth—sort of like a shorter, leaner, more Jewish Norman Mailer. When my brother and I were young, Ben's favorite thing to do was to stick his head out of a doorway or around the corner of a wall, and then grab it back with his own hand, screaming for help as he did it. He also coined the phrase "No coughing allowed," which became a standard of the Siegel-Kaplan lexicon. Why it never caught on with the population at large is still a mystery.

Ben had a way with language and had been a comedy writer for the radio stars of the thirties—Jack Perl, who was known as Baron Munchausen; Joe Penny; the Marx Brothers, and Burns and Allen. He also had a job writing lyrics and poems for the Quality Art Novelty Company. He sat around with a bunch of WPA writers, one of whom was Robert Benchley, and wrote unprintable poems that started "Roses are red, violets are blue . . ."

There was nothing better than a visit with my grandfather.

We played endless games of ghost, tic-tac-toe, and dot, which is the game where you connect a grid of dots, one line at a time, trying to make boxes, which you then write your initial in. The one with the most boxes at the end wins. This is a *very* good game. We also walked South Ocean Boulevard looking for fallen coconuts so Ben could make a coconut shake. When we found one, we would take it to the side of his building's garage and smash it repeatedly against the concrete wall until it cracked. We played together in shuffleboard tournaments. Ben was a South Florida champion in his age category and had a considerable collection of paperweight trophies. They were not exquisite and so were relegated to the lower shelf of his night table and the powder-room medicine chest.

Ben was always up for something, or up to something. He hummed and whistled constantly, little doodly nothings, snatches of old-time songs, vaudeville numbers; there was a funny sly quality to it, like he was standing lookout for the big heist. He picked out songs like "The Michigan Rag" on our piano, running both hands up and down the keyboard, even though he didn't really know how to play. He told jokes and wrote poems (yes, "What's Gnu at the Zoo" was his) and he wrote fantastical letters addressed to My Li'l Cin and signed Dit Loves Cin, written in a heart with an arrow through it. My brother and I never called him Grandpa, or Pops, or Pop Pop, or any traditional endearment; we called him Dit, because when my brother was a baby he heard my mother call him Dad, and Dit was the best he could do and it stuck. My heart still aches to see that configuration of letters.

Once, when Ben was visiting us in Connecticut, I complimented him on his sport shirt. That night I found the shirt folded on my bed with a note that read, *You can always have the shirt off my back.*

He died the year after I graduated from college. We flew down to Florida and stayed in the same hotel as Aerosmith. I cried everywhere we went. I was crying in a corner at the funeral home when I suddenly heard the familiar whistling and humming. I turned sharply to find a Not Quite Ben, a Faux Ben, who was in fact my grandfather's brother, Frank, whom I'd met maybe once. They looked alike, and obviously they'd had the same noisy habit. He trailed me the entire funeral, a puckish apparition, and I know it was ungenerous of me but by the end I wanted to kill him.

During the service Lil, who loved a grand gesture, threw her arms around the casket.

After Ben's death, I occasionally flew down to Florida by myself to see Lil. As soon as I arrived I would change into a bathing suit or shorts and a T-shirt, and we would go sit by the pool. She would show me off to whichever friends were not across the boulevard at a spa called Le Mer, or, as they called it, *the* Le Mer, which, for the benefit of those who do not speak French, means The The Sea. Lil always had something to say about my summer clothes. No shorts were ever short enough. No midriff sufficiently bare. "It's what all the young people are wearing," she would say. In her insular South Florida world, that is, her building, "young people" could only mean the succession of foxy Jamaican girls who shopped and cooked and kept house for her. She loved these girls, loved their beauty and their youth, and they in turn loved my grandmother's high spirits and her generosity. Lil also had a fondness for voodoo, and together they indulged in their mutual superstitions. Lil even read tarot, although, ever genial, she had pulled the death card from the deck so it would not come up in her readings. One afternoon, we went across the street to the Diplomat, where a recently deceased friend's daughter was staying, to pay a sort of belated shiva call. When we left, my grandmother stopped in front of the elevator. She said: "That woman gave you the evil eye. Turn around three times and then spit three times." I turned around. "Puh puh puh," I said. "Puh puh puh," she said. "Okay, now we can go."

Besides her jewelry, there was one thing that Lil really wanted me to have. Or to do. She knew I wanted to be an actress and she thought that if only someone could get in touch with Nat, I would be set on my way to international stardom. Nat was my grandmother's uncle, Nathan Birnbaum, or, as he was known to the world, George Burns.

"You must write to Nat," my grandmother said every time she saw me.

"Why would he read a letter from me?" I asked.

"You just tell him that you are Lily's granddaughter, his sister Annie's great-granddaughter."

Lil's idea of help probably included a seven-year studio contract with Warner Brothers (if they were still giving those out) or at least a screen test for the next God movie. Maybe George would introduce me to Brooke Shields. Or get me an agent. Even Lil knew a person had to have an agent, although she was probably thinking along the lines of the agent character Red Buttons played in a terrible 1960s biopic about Jean Harlow. According to the movie, Red discovered Jean trying to filch a drumstick from catering on a film set she'd crashed. He was convinced that she had "it" and would someday make them both rich. So he took her on. He drove her from audition to audition. He introduced her to lecherous stars and producers. But he was not a pimp. He just believed in her and that's how you got known in those days, scampering about in a bathing suit at a famous person's pool party. Maybe this was why Lil was always after me to show a little more skin.

"You know that Nat discovered Ann-Margret?" said Lil.

After several years someone in my family finally coughed up a number. I spoke to George's secretary, who had been with George for decades and whose name may or may not have been Jack. I explained my pedigree to Jack and what I wanted, which was just to meet George, that's all. Lil gave me two pictures to bring with me. One was a formal of her mother and father and the other was of George, Maurice

Chevalier, Al Jolson, and another man, their arms slung around each other's necks, laughing. It was a wonderful picture. Even though it was fifty years old, I felt a pang that no such glorious picture would ever be taken of me. I often experience a sense of loss for having been excluded from events that took place decades, if not centuries, before my birth. I still have a crush on Joseph Cotten.

So I flew to L.A., ostensibly to see friends and have "meetings," but really to see George.

Jack met me at the door of George's office, which probably had not been redecorated since 1955. I had a half an hour, said Jack, after which George was scheduled to do a radio interview. George was seated in one of several tall director's chairs. He did not look a day over ninety-gazillion. Once again, I explained who I was. I brought out the pictures and showed them to George.

"This is who?" George asked.

"Your sister."

"My sister."

"Yes, your sister Annie."

"My sister Annie."

I realized then that I could be anyone, any scheming young wanna-be starlet, waving a couple of raw-edged, sepia-toned photographs in his face, claiming to be his blood relation. If only I *was* a scheming young wanna-be starlet, I'd probably have gotten a lot farther. I put the pictures back in my purse and fell silent. George told fifteen jokes in a row.

Then he asked me if I had an agent. No, I said. He yelled to Jack to get me a meeting with a big agent friend of his, a man he had had lunch with every week for the past forty years at the Hillcrest Country Club. Then Jack came in to say that it was time to call the radio station. George wished me well, got up, or rather down—the chair was high, he was short—went to his desk, picked up the phone, and started telling jokes.

I had my meeting with the agent, an exercise in futility. He seemed as stymied by my presence as George.

Lil was philosophical about the encounter.

"He's as old as the hills," she said.

On one of the last visits I had with my grandmother before she died, we spread her things out on the bed, and among them were some I had not seen before. A small green chiffon scarf and a jeweled pillbox.

"These belonged to Gracie Allen. They're for you," Lil told me.

"Well, at least that's something," I said.

"She was a very sweet woman. George loved her more than life itself. He visits her every day in the mausoleum."

Obviously we had seen the same Barbara Walters interview.

"Your mother has Gracie's pocketbook. You'll get that too, one day, puh puh puh." She spit through a V in her fingers.

Then she showed me a gold locket.

"This is from England. It's exquisite," she said.

On the front was the imprint of a bird and on the back a spray of flowers. It *was* exquisite. I opened it expecting faded photographs of my ancestors but found instead a crusty green residue.

"You must never wash it out," said Lil.

"What is it?" I asked.

"It is a special love potion and it will bring you good luck."

One of her island girls had given it to her, and a year later, when my mother and I were going through Lil's things, I found its source, a small jar with an ancient label that read COMPELLING POWDER.

Ah, my inheritance.

the story of r

Part One

How do you tell the story of someone you know, whom others know, when that story ends badly, infamously, even ignobly? How long do you wait before its telling is neither a betrayal nor a humorous anecdote, despite its inherent and obvious ironies (despite the fact that it actually *is* funny), but is rather a reasonably considered part of the fabric of your *own* story and therefore demands to be told, that is, if someone were apt to make such a demand, like your publisher who is paying you for a book of personal essays?

All right, so my therapist went nuts. There, I said it. Her life fell apart and she crinkled and crumbled slowly before my eyes until I knew, for sure, that I was cured.

* * *

She'd always been odd. She'd always been messy and bohemian, an earth mother in loose triangular velvet dresses and elaborate beaded necklaces. She wore large, buggy glasses and had long center-parted mussed hair. Her office was crammed with classic therapist voodoo: a large and valuable photograph of a famous poet on one wall, a Tibetan blanket on another, two corduroy club chairs, a couch, plenty of books, and a vast collection of exotic-looking figurines that in time came to include those small rubber promotional dolls that come with Happy Meals at McDonald's. But every week more *stuff* arrived and over the three years I was with her the condition of her office went from homey-messy to creepy-filthy. The place became a dumping ground of her personal and professional detritus. A computer that didn't work, old blankets, lamps, piles of books and magazines, stuffed animals, real animals. Some time in my first year a dog appeared, a poodle-ish affair, and then later a cat, too, and then another cat, a skinny, icky cat. She once asked me if I wanted to buy this big black and white photograph of an orchard that she knew I liked. A patient had given it to her in lieu of payment. I considered it for several weeks until one day I saw it propped between the waiting-room sofa and wall with a lump of what could only have been cat shit on it.

If ever a room was the manifestation of a troubled soul, this was it. If ever there was a textbook case of impending emotional disturbance, this was it. And she frequently took mysterious phone calls. She popped pills. Sometimes she fell asleep with her mouth open. I could go on and on, but what's the point? The party was over. I was on my own. And still, I think she helped me. By the time I left I felt pretty together. Everything's relative.

I'd come to her because my idiot boyfriend had broken up with me so meanly, so damagingly, that in a brief moment of insight following weeks of despair (yes, it hurts to be dumped, even by a fool. That could be the title of a book— *Yes! It Hurts to Be Dumped, Even by a Fool!*) it occurred to me that I didn't want to devote my entire year to getting over him. Also, after playing two pathetic though not entirely unironic songs I'd written about him to my friends John and Sarah (not their real names), they wisely suggested I call a therapist—in fact, theirs.

Rule number one: Never go to a therapist your friends go to.

This would be the first time I had sought the advice of a professional. I'd been so sick to death of hearing my friends quote their shrinks that I'd vowed to stay mentally ill for as long as possible. But, as they say, time marches on, or is a winged something, so I promptly left a teary message for R, as I will call her. Nothing inspires tears so much as admitting to a total stranger's answering machine that you need help. I proceeded to see her once a week for the next three and a half years. Here was my diagnosis, for insurance purposes: DSM III–R: Axis I 300.02, Axis II 301.50. Look it up if you want. Our sessions were fairly cerebral; R was not a Freudian, whatever that means, but more of a Jungian, whatever that means, and I never became a fetus or pretended to be my mother or a tree or a wild animal. And I only cried when I arrived at her office immediately after witnessing something sad, as I did the day I crossed paths with a crippled young man on crutches, inching along to wherever, slug-slow but determined, just outside R's building. My capacity for self-pity is exceeded only by my capacity to pity others, and R and I went to work on both.

We talked about my self-image, my family, my work. We set little goals. *Could I wear some beads maybe or a sort of sexy shirt sometime?* I liked R because she never said, "How did that make you feel?" to me and because we took time to discuss books and current events, like real people. Actually, now that I mention them, we also discussed Real People, as in people we mutually knew, some of whom were her patients. R hated the woman who had the appointment before me and she often spent the first fifteen minutes of my appointment venting. My hints to change the subject to something more relevant, i.e. *me,* went unheeded. One day R was positively incensed that the woman had criticized R's increasingly unhygienic surroundings, suggesting that maybe R give the place a coat of paint or have the floors fixed. At the time I sided with R in her righteous rage, but I secretly toyed with the idea of giving my co-patient a high five next time I passed her in the hall. The floor was, in fact, buckling from a small flood some months before and had not been fixed. The walls were marked up and the paint was peeling in spots.

Occasionally, R spent time updating me on the progress, or lack thereof, of a thirteen-year-old anorexic and her controlling, warring parents. Or, on more familiar ground, she liked to chat about which among my friends/her patients were talented and which weren't. I didn't mind this so much, especially since we always came to the conclusion that I was exceptionally talented. (R was writing a play, so she said, and was always casting it and recasting it within our small theatrical circle; she liked being one of us, us creative folk, such as we are.) It was the breaches of confidence I found most unsettling, but I didn't have the guts to ask R not to make them. They were always prefaced with "I know he wouldn't mind you knowing this" or "I'm sure they are going to tell you" or even, "I think they would *want* me to tell you," which is how I found out that one friend's boyfriend had a severe drinking problem and two other friends, whom R saw individually and as a couple (isn't there a rule about that?) were having marital difficulties and were in fact separating.

Rule number two: Be careful what you tell your therapist.

Still, I got along with R because I felt I was getting help without getting my guts vacuumed out. I'd had a boyfriend who did that. If I was unhappy with him he'd flip the thing around and force me to search my*self* for the source of my displeasure and together, with his urging, we'd scrape and scrape and suck and suck until I felt my vital organs rising up in my gorge. I'd eventually start to cry—could you blame me?—and he'd feel vindicated. We wouldn't even have gotten at any truths, since the main truth was that he was an asshole. We would just hypothesize me into a coma. Anyway, since R didn't completely inspire my trust, I kept certain things from her, which was empowering in itself. Sometimes, before I told her something private, a little voice in the back of my head asked me if I wanted my friends to know this thing as well. Good practice for later in life, I should think. Thanks, R.

One day, R asked me for one of my migraine pills. She had a terrible migraine and was out of medication and couldn't reach her doctor. Now, it just so happens that I don't really take a designated migraine medication for my migraines. I take Tylenol 3, or for the uninitiated, Tylenol with codeine, which is a narcotic. It doesn't stop the headache itself but it almost always kills the pain. Okay, let's pause here for a moment and discuss narcotics. I am particularly partial to the combination of pain relief and seemingly paradoxical, but not if you know your narcotics, sleep- slash buzz-inducing element, where you lie down and talk your head off until you suddenly, exquisitely pass out. If you don't take a lot of them, because they are addictive and constipating, they are your friend. If you take a lot of them, they become your friend in that way a woman who is sleeping with your boyfriend becomes your friend—you want to get rid of her but you can't no matter how hard you try. That's just a theory because I'm not addicted, although I have experienced the latter, which is a *nachtmare*. That said, I gave her one.

R offered to pay me for the pill, which seemed ridiculous at the time, maybe even illegal, and I refused to let her give me back any of my hard-earned money. Over the years, in addition to popping "aspirin," or something, throughout our sessions, she continued to "borrow" my Tylenol 3. There were weekends she left messages for me at home, asking if I could leave some downstairs with my doorman for her, she couldn't get in touch with her doctor, or she didn't have a doctor, or she was between doctors. Or she lost her wallet or her dog ate her meds. She actually used the wallet excuse several times but not the dog one, although that dog certainly looked strung out to me. My current boyfriend found the whole business entirely improper. But I felt strange refusing her so sometimes I just didn't tell him.

Rule number three: Don't share your prescription medication with your therapist.

A long time ago, my friend Kate wrote a performance piece about how shocked she was when she realized her therapist was seven months pregnant. She knew so little about the woman and the sessions so focused on her, she simply failed to notice. I knew that R was married, a second marriage, and that there was a college-age daughter from the first and a son from the second. I only once saw R's husband, although he often showed up during my sessions and R would excuse herself and meet him in the waiting room. From what I understood, which, as usual, was more than I should have, he took care of their son a good bit of the time. If he didn't actually swing by, he would call R on the phone at least once, sometimes twice a session, reporting their whereabouts and planning the next time he would call in. I didn't know what to make of this, and it was pretty annoying to have to stop cold in the middle of an earth-moving epiphany so she could confirm her son was at the playground or the library or a birthday party. She would say, "Okay, where are you going? And what time are you going to be there? Okay"—she would look at her watch—"call me when you get there." And then twenty minutes later the phone would ring. Besides making me privy to these strange reconnaissances, R would also talk about how bright and talented her son was—at six, a natural actor.

I am not exactly sure when, but at some point during the third year, I began to get the impression that R was unraveling. She popped more pills and made more phone calls, sometimes asking me to wait in the waiting room, which was getting almost too disgusting to sit down in. Magazines and newspapers were piling up and animal hair was everywhere. And her office looked like the spot under a tree where you heap your belongings after your trailer has been overturned by a twister. I didn't want to touch anything there or have it touch me. I felt hemmed in, claustrophobic, during our sessions. The place had acquired a new mustiness. No longer content to be merely dank, as it had always been, it had become dank's evil twin, rank. I tried to maintain a forward focus, so as not to get grossed out by something I might notice on the table next to me or on the floor. As I remember, R once claimed to have an office downtown. Could it be possible that this nightmare was replicated somewhere else? The subject of R's own apartment, which it seems was in flux and which was why, ostensibly, so much shit had found its way here, came up during one session and then became a staple of subsequent ones. She told a roundabout and ever-changing story about money paid to a landlord before a promised renovation, some sort of swindle, a safety-code violation or mismanagement—I still have no idea exactly what happened. It didn't dawn on me that she may actually not *have* an apartment until one day when I saw into the other room in the office, the door of which I had never seen open, and witnessed what could only have been a storage job subcontracted to the Beverly Hillbillies. Sometimes I still think it was a vision I saw in a dream—furniture piled to the ceiling, clothing and suitcases and maybe even cats. R occasionally mentioned that she and her six-year-old son had spent the night, "for fun," at the office. Yikes.

During one session R actually fell asleep, sitting up, while I was talking. Her eyes got heavy and her mouth became lax, so much so that a gleam of saliva appeared on her lower lip and I was terrified she was going to drool. Finally her head dropped to her chest. I stopped talking, stared at her in amazement, feeling like I was in a movie, and then said her name sharply. She bolted up, mumbled an apology, and we continued. Some instinct told me not to try to discuss what had just happened. Can you imagine anything scarier than opening your therapist's Pandora's Box? And then, with a shock, I realized it was already open and I was sitting in it.

Obviously, something was going on, had been going on, with R for some time. It was not like I hadn't noticed, it was just that I was more interested in how *I* felt than how *she* felt. That is how it is with doctors. We go to them because we have a problem and it is *very* inconvenient if they develop one that is bigger than ours. Best to just ignore it as long as possible. Which is what I did, and may I say not for nearly as long as some others, but I'll get to that later. I am more assertive now than I was then (thanks, R) and these days I wouldn't stand for any of that nonsense. But six or seven years ago, well, I was in transition and I didn't want to have to start all over with a new therapist.

Rule number four: Always leave before the party's over.

Especially when the party moves to a hotel room.

One day R called to tell me that she was moving offices and would, in the interim, be seeing patients in the living room of a suite at a hotel near Fifty-seventh Street. It all seemed perfectly respectable, yet I was suspicious. Where was all her shit? The hotel room was immaculate—just me, R, the requisite chairs and curtains, a desk, a coffee table. That's when I knew I had to get out while the getting was good. R looked fidgety and ill at ease. Like a dog on a waxed floor, she couldn't get a grip. While I was there some food arrived; R often ordered food during my sessions and then ate, messily. Then her son, who I didn't know was there but was in the bedroom, spilled some water while jumping up and down on the bed and called, guiltily, for his mother to come. I felt bad for him. It was the middle of the day. Why wasn't he in school?

What little was left of the session after these interruptions was devoted to me, but still, I had to think up things to say. I found R and her mess so distracting that I just didn't feel that much like talking to her. She had become the friend you no longer like enough to confide in so you feed them little drips and drops of your life. Just enough so you don't have to go through the bother of actually ending the friendship. Also, she was so obviously worse off than I that it would have been like telling a starving child you've got a hankering for a cup-cake. I always felt relieved when I had an excuse to skip a session, like a trip or an audition or if I was sick (a migraine, hooray!). All spring I kept trying to figure out how and when to end it.

I finally got the break I was looking for when I was invited to be in an intensive and time-consuming workshop at a New York theater. I paid one last *visite à l'hotel* and at the end of the hour I told R that I was feeling strong and happy, that I wouldn't have time to see her for a while, and perhaps this would be a good opportunity to take a hiatus. She did not rejoice in my wellness. In fact she was pretty nasty about it. I told her I'd call her when the workshop ended, but I left secretly hoping never to see her again.

I didn't. Which leads me to Part Two.

Part Two

Three months and a few extra weeks of procrastination later, I telephoned R, only to find her number had been disconnected with no forwarding information available. I was mostly relieved but vaguely alarmed, although not enough to pursue it, or rather, her. I was curious, though, and a few friends slash fellow patients confirmed that while R had fallen off the radar, she had not disappeared entirely. Evidently, she had so insinuated herself in the lives of several of her patients that not only did they continue to see her, but they allowed themselves to become entangled in her own disastrous affairs.

Here's what I know. Actually, for legal reasons, let me revise that. Here's what I heard: R borrowed a total of approximately eighteen thousand dollars from three of her patients. One would have to surmise she borrowed more than that from patients I did not know, presuming there were any. She had neither an apartment nor an office and moved from hotel to hotel, including, at one point, the Helmsley Palace. Sometimes she had her husband call patients to ask for money or groceries. One night they asked a patient, a friend I'll call Howard, to pay their hotel bill so they would not be evicted. Another night R asked a young woman who had already lent her thousands of dollars to bring food to a hotel, claiming that she and her son had nothing to eat. When the woman asked R—through the closed door of the hotel room, as R would not open it— about being paid back, R threatened to call the police if the woman did not stop harassing her. Another patient I knew who was particularly dependent on and devoted to R, let's call him Mark, not only loaned her money but became actively involved in an effort to keep her and her family afloat. She preyed on his vulnerability, on his good-hearted-ness, entreating him to not let her "fall through the cracks" as so many do when they become homeless. On the face of it, this is not an unreasonable request. People get into trouble sometimes, their lives succumb to some sort of fatal disarray and they need help. Well, this wasn't that. Had it been, from all accounts the aid she received from her patients should have buoyed her for some time. Rather, the money was obviously disappearing as fast as it was coming in. When Mark was tapped out both financially and emotionally and had finally cut himself loose from R, he brought her up on malpractice charges with the appropriate body politic, the Office of Professional Discipline. In the course of this process, he actually discovered that R was not, at present, properly registered as a social worker. She had her credentials, she was licensed, but she had not kept up with her registration fees for years. Unfortunately, the Office of Professional Discipline so bungled the case that it is still unresolved. Something, Mark tells me, about their process server not dating a summons properly.

I wanted more information than seemed fair torturing Mark for, so I called the Office of Professional Discipline. I identified myself as a journalist (why shouldn't I?) but after giving them R's name was told they could not confirm there was any action pending against her. "But you are not denying it?" I asked. The woman understood what I wanted. (*If I sneeze twice and you don't say gesundheit . . .*) She went to talk to her supervisor and then came back and repeated that she could not *confirm* it. That was good enough for me. Besides, even though I didn't learn anything, it was fun to say I was a journalist.

I called Howard. He was still being treated by R through a good bit of this, and he confronted her several times regarding the money she had borrowed from him and from fellow patients. She told him variously that the money was not a loan but a gift and that she was not expected to give it back, that the money was an advance on future sessions, and that it was owed her because for years she had undercharged her patients. Howard's own therapy finally ended when the check R gave him to repay him for paying the hotel bill bounced. He never saw her again after that but he kept up with her by e-mail for quite some time. He was worried about her and although he was annoyed to hear from her that she had bought a PowerBook with money borrowed from her patients, he continued to try to help her. In response to his queries about her well-being he received lengthy rants detailing her precarious circumstances. Sometimes R wrote messages all in caps that would trail off into non sequiturs or end abruptly, midword, as if she had just passed out at the keyboard.

What can we take away from all of this, I wonder? I spoke to Howard again recently and he told me a hair-raising saga about how R once encouraged, if not abetted, a romance between himself and another patient, only to destroy it when it blossomed by revealing each patient's confidences to the other. I find that the years have not mellowed his anger and confusion. John and Sarah, the couple who had originally recommended R to me, will always wonder how much she may or may not have manipulated the ups and downs of their marriage to suit her own purposes. And there must be countless others whose stories I do not know. My boyfriend heard a strange but not unlikely tale from his own therapist about a patient of R's who may have, as it turns out, unwisely, loaned R her credit card.

God only knows what has happened to her young son.

*   *   *

As for myself, well, I am philosophical. I liked R quite a bit in those early years. She commiserated, which is something I needed, still need, perhaps more than therapy. Justification as Cure, that's me. You will remember that I didn't much like the idea of being in therapy in the first place. Considering R was not properly registered, I think it stands to reason that if all I did was pay a crazy lady seventy-five bucks a week to chat with me about myself, it may have been stupid, but it wasn't therapy. Hey, look at that, Justification as Cure really works.

Who was R? A lost soul or a manipulative user? Or both? A dedicated healer or a quacksalver? (Means *charlatan*—I found it in the computer thesaurus.) Perhaps all those phone calls she made and took, which I'd presumed from R's discreet one-or two-word responses to be patient emergencies, were actually drug communiqués. It is impossible to know. Whatever the case, I am sad and sorry about her dissolution. She helped me through some tough times by laughing with me at my oppressors, which gave me validation and built my confidence. I am a stronger, though not necessarily a nicer person for it, but I am working on that last part. Sometimes, now, when I am low, when I feel misunderstood or ill used, I think of what R might have said to me, had I arrived for a session thus. We would have had a little therapy, talked about the new Alice Munro collection, and then maybe together we would have subtly put down a colleague or friend of mine, just enough to lift me out of my slump. And, of course, she would have suggested I put on a pair of dangly earrings. When R was good, she was very good. And when she was bad, well, she was sort of good, too.

waiting

———

Even if you are a waiter for a very short time, you are doomed to have waiter nightmares for the rest of your life. You go into work and your uniform is missing or you can't figure out how the tables are numbered or you've suddenly developed a limp. Your dead aunt Rose, who was always impossible to please, pops up at one table or someone you made fun of at camp or an ax murderer is demanding their appetizer. You miss your wedding because no one will cover for you.

I was a waitress for almost four years in a restaurant in Soho I will call Mariella. Mariella was owned by a despot named George, who, when he was through screaming at us at staff meetings for our various and sundry infractions—spotted aprons and askew table settings and snide attitudes and tardiness and wastefulness and our overall failure as viable human beings—would then make the startling pronouncement that the greatest achievement of our lives was that we were working for him. There was some truth to this statement. Without a doubt, George was the most loathsome person I had ever met, and if somehow one could manage to avoid his wrath and last more than two weeks on the wait staff at Mariella, that was, indeed, an accomplishment.

There was actually a "Mariella," for whom the restaurant was named. She was a tall, thin, very attractive Jamaican woman known mostly for wearing tiny sparkly dresses, spike heels, and outrageous wigs. Mariella could be both funny and brutal—to call her capricious would be too flattering—and you remained in her good graces, or not, depending on whether you were a) keeping her amused and b) keeping people's wineglasses full and dishes cleared. If they were going to drink, they could stay; if they weren't going to drink, they had to leave, because the restaurant made most of its money on liquor. Or if they were having no appetizers, or appetizers as main courses, or splitting main courses, they had to leave. Or rather, *you* had to get them to leave by continually asking them if there was anything else they needed and then clearing their table until there was literally nothing left on it but the tablecloth. You couldn't just put down the check, because that would have been rude. Mariella was also a relentless publicity hound, and you could count on a reasonably pleasant lunch shift if she had appeared that morning in a chef's hat and stilettos to make fried chicken during the food segment of some morning show, or if her name had turned up on Page Six of the *Post*.

It is not as if there is a writ or edict, something on file at City Hall declaring that all struggling actors take jobs as waiters with no regard for the fact that perhaps their backgrounds or educations recommend them, in the opinion of their parents, to a more lofty purpose. Mainly, I needed a job that would have some flexibility so I could go to auditions and rehearsals. I also needed a job that would not require that I wear stockings, which I hate. Waitressing at a salad bar, as I had done in college, and which I believe is an oxymoron, unfortunately did not qualify as restaurant experience. However, as a testimonial to my acting school, at my interview I acted the part of a waiter and was hired.

Mariella was, at the time, a very popular restaurant and therefore a very desirable place to work. It offered expensive Jamaican-ish food and was famous for its two deceptively strong frozen rum-and-fruit drinks, a red one called a Mariella and a blue one called a Montego. I would come to know these drinks well. I would serve several of them to a person and perhaps by the end of the meal the person would serve them back to me in a slightly altered form, most likely in the company of some half-digested onion rings or curried chicken or pecan pumpkin pie.

The restaurant was also renowned for its raucous Gospel

Sunday Brunch—perhaps the most dreaded wait shift in all of Manhattan. We would arrive at nine-thirty in the morning, cut a thousand bagels, and move all the tables and chairs around in order to fit 233 people, the exact number permitted according to the fire department's Maximum Allowance sign. By eleven o'clock, the first clamoring horde—you'd think they hadn't seen a bagel since the Carter Administration—was stuffing its collective face with challah French toast and Montegos while singing eighty-three verses of "This Little Light of Mine" along with an overmiked gospel group. Actually, the cacophony was produced by two main elements: the sizable black contingent giving it up for Christ and the equally sizable Jewish contingent screaming for their food. And, as always, Mariella would be following us around with the full ashtrays she'd picked up off of our tables, shrieking, "What is this? What is this?" Just when it seemed like All God's Chillun had been fed and/or saved, the bagel cutting began again in preparation for the second seating. Sometimes while I was cutting bagels I would think about how once my grandmother told me that as a young woman she was courted by one of the Lender brothers. If she'd married him we'd all be rich now and I wouldn't be here, cutting bagels, and even if I were here, Lender's bagels are pre-cut.

The brunch shift would finally end around six o'clock, just as the waning sunlight signaled the official demise of the weekend.

* * *

I adapted well at Mariella under the tutelage of ten or so actor/waiter/homosexuals, who taught me how to open a bottle of wine with a pocket corkscrew, stack plates up my arms, and make cappuccinos. They also, in accordance with the laws of their people, taught me the original choreography from every Broadway show of the past fifty years, including *Broadway Babies of 1925*. In college, none of the people I knew to be gay were "out." If they were not completely assimilated into the mainstream they were relegated to the small though vocal Lesbians and Gays at Penn, or LGAP, pronounced *el-gap*. I didn't know anyone in LGAP. I'd had no gay friends in high school either; people in suburban Connecticut in the seventies weren't allowed to raise gay children. But at Mariella, it was as if a dozen Rosalind Russells were starring in a never-ending Jacqueline Susann movie. The banter was smart and snappy, but it was all about sex and drugs and show business. And management. The general consensus was that the former were good and the latter was bad.

Unfortunately, I didn't do drugs and there was only one straight waiter at Mariella, Pete, and he had a girlfriend. Although that did not stop us from smashing ourselves together one night on the street outside an after-hours bar where we and eight or so of our homosexual brethren had adjourned for a post-shift sousing. That night Pete and I were drawn together in that age-old, time-tested fashion called the Process of Elimination.

Sometime during my first summer at Mariella an English guy named Matt showed up and promptly became the object of one of my three or four life's great obsessions. An attractive man with an English accent is wonderful thing. Even the Boys wanted a piece of the action. Matt and I liked each other a lot very quickly and almost had something going and then everyone started teasing us and we became very self-conscious and nothing happened, which, of course, ensured him a place in my psyche until the twelfth of never. At the end of the summer Matt returned to London. I actually hunted the guy down in his sister's Shepherd's Bush restaurant a year later, while pretending I had come to London to visit a friend. He took me driving around on a motorcycle and we made out like fiends and then he left on holiday with his mates. A year after that, I was working the bar tables at Mariella and looked out the double glass doors and he was standing there holding a bicycle. I went outside and he pushed me against the building wall and we kissed like we were dying and then he pedaled away to go on holiday with his mates. I guess we weren't going to get married.

Over the years waiters, managers, and chefs came and went with startling frequency, a reflection of the whims of the owners, of the AIDS epidemic, and of show business. When the national tour of *Beauty and the Beast on Ice* calls, you go. But despite the ever-revolving door, there was an almost instant camaraderie among us. We were quick to accept new staff, providing they were competent, knew the lyrics to at least five Sondheim songs, and could survive a mandatory hazing period which consisted of being sent into the basement of the restaurant swathed in garbage bags to clean the pestilent onion-ring-batter machine, even though such a machine did not exist. "Keep looking," we would call down. It also helped if you could speak with a foreign accent. Doing Meryl Streep speaking with a foreign accent was particularly impressive.

Every night we put on our waiter costumes and our efficient-waiter smiles and we went out there and tried to make people happy enough to tip us 20 percent. And between taking orders, delivering food, and avoiding Mariella, who nightly could be found teetering about the restaurant alternatingly sucking up to celebrities and venting her spleen, we convened in the waiters' pantry and made fun of everyone and everything in sight. There was a highly entertaining, though short-lived, staff newsletter called *Eat the Press*, chronicling the exploits of various waiters and managers and customers. It was a saucy publication and after two volumes was quashed by our humorless employers. Being a waiter at Mariella was as close as I have ever come to fulfilling the obligations of my vaguely socialist heritage. Isn't that what young socialists did? Take orders from the bourgeoisie and then gather in tetchy clumps to make coffee and compose propaganda? At Mariella, I was among the proletariat, a worker bee. Insurgent. Or maybe just insolent.

* * *

Here is how to be a good waiter, which means getting large tips: always give an opinion when asked. The curry is better than the lamb chops. The snapper is so-so but the halibut is delicious. People love when you tell them not to have something. It inspires trust. Shake your head conspiratorially when they ask about the osso buco. That's it. The rest is common sense. Be nice but not intrusive, be relaxed but let them know you're in control. Get them stinking drunk.

One of the many pitfalls of working in a restaurant is that eventually you will wait on your peers. Or people who had been your peers before they became successful bankers and you became a waiter. First, there will be the requisite "Hi!" "Hi!" "How are you!?" "What are you doing?!" (What am I *doing?* I think it's pretty clear: Good evening, my parents spent fifty thousand dollars on my education, would you like some more bread?) After the initial pleasantries are dispensed with, you will embarrass everyone with "Let me tell you about our specials." Hopefully the food runner will deliver the food so you save yourself the agony of "Enjoy your meal." You must, however, check up on them, refill their water glasses, and, noticing their empty wineglasses, ask, "Another bottle?," to which they will sheepishly shake their heads no, sorry not to be spending more money at your table. The episode will deteriorate at a fairly even pace, through dessert and the proffering of the check, until, finally, during the good-byes, someone enthusiastically pronounces you *Neat!* for being stalwart enough to wait tables while pursuing your pathetic dream.

Another pitfall is waiting on celebrities, which is inevitable if you work in a "happening" New York restaurant. It is very hard to strike the right note between *I know who you are* and *I don't care who you are*. And they, in turn, are either apologetic for being celebrities and fall over themselves proving they are not jerks, or they *are* jerks.

Most often, though, you will wait on people who expect you to contribute to their overall happiness and well-being, or make them forget about their crappy day, or their stinky marriage, or the fact that their mothers didn't love them enough, and who will behave badly if you disappoint them or will behave badly because that is all they know how to do. At Mariella, in the eighties, most of the customers were either hyped-up Wall Streeters flailing money, or neighborhood book publishers on long lunches because enemies of Salman Rushdie were threatening to blow up their offices, or dissatisfied women with recent face-lifts. I could always tell when a woman had had a face-lift because when I came to the table to take her order she'd look surprised to see me, as if even though I was obviously her waitress she didn't think I'd show up. She'd look surprised when I brought her a drink, as if she'd forgotten she'd ordered it. She'd look surprised when the meal came. Is that for me? What did I order? Surprise, you ordered the tuna.

But the money was addictive. I came home most nights with a pile of cash, which, being me, I dutifully put in the bank. The job saw me through my last year of acting school and several low-paying theatrical productions, one with the very unfortunate title *The Little Planet of the Heart Is Vast.* It's not.

I worked Christmas Eve, New Year's Eve, Easter Sunday. I worked Valentine's Day, where people would propose to one another over corn and crab fritters, or get drunk and make out at the table or in the bathroom downstairs or sometimes just sit in sulky silence, waiting for me to bring their food and give them something to talk about. But I did not mind working the big holidays. It was extremely convenient having an excuse for why I couldn't go to holiday parties where you were supposed to wear all-white clothing, or why I didn't have a date for Valentine's Day. And there was comfort in knowing that at midnight on New Year's Eve I would not be standing half drunk and headachy in a crowd of people who were obviously *best best* friends with everyone there but me, wishing I was home in bed.

The tough shift to work was Sunday brunch. That was the day ordinary people, happy people, rolled out of bed and read the paper over breakfasts of tea and croissants and jam and played co-ed touch football in the park with their friends and friends of friends, and if they did not already have a boyfriend, they found one on Sundays. Or they were away for the entire weekend cavorting at the beach or admiring the autumn leaves or wedeling (if anyone still does that) down the slopes until the sun slipped behind the tree line and then they would drink and dance in their ski boots to a guy playing "American Pie" on a guitar, and then, *then,* they would get into a giant outdoor hot tub and tell hilarious anecdotes about the day's adventures and touch one another beneath the black bubbles. *This* was the life of a non-waiter. *This* was the life that was going on every Sunday, the life anyone who wasn't cutting bagels for the teeming multitude could have. It was on Sundays that I was stung by the loss, regretful, feeling that my real life hadn't yet begun, and wondering when it would.

On the night I turned thirty, about ten of my friends met me at Carmine's for dinner. I sat at the center of the long table under the misguided impression that it would make me the center of attention, only to discover that I was unable to be part of the conversations that had sprung up on either end. The man whom I was vaguely dating stood me up, and I was forced to add another layer of false jollity on top of what was already a mille-feuille of pretense. I drank too much red wine in the hope that it might inspire one honest emotion. It did: relief. I could have been waiting on my table rather than eating at it.

world peace

———

LAST night I slept for forty-five minutes. I fell asleep at 3:34
A.M. and distinctly remember looking at the clock as I awoke
at 4:19 A.M. An investigation this morning conducted by
myself and my cousin, Erica, who is staying with me in my
apartment, revealed the cause of my insomnia to be an unin-
tentional overdose of the nonprescription medication Maximum Strength Multi-Symptom Midol.

It all began when I was invited to an ice-hockey game by a
man who on our first date had, without provocation,
explained that discussions of menstruation, or the P-word, as
he called it, did not particularly *interest* him. Why this tiny
nugget of information did not register a warning in some
appropriate quadrant of my brain, I cannot say. All I know is

that I was sufficiently impressed with the fit of his Levi's to put a handful of Maximum Strength Multi-Symptom Midol into my pocket before I met him at Madison Square Garden. When I got there I went to a water fountain and took two, as a preventive measure.

Happily, my little friend, as someone's mother, thank God not mine, once called it, failed to appear. As luck would have it, however, I was afflicted instead with a migraine headache, another source of mind-altering agony sure to enhance the enjoyment of any date, particularly one taking place in a sports arena. A man always remembers you when he has to take you home at the top of the third period, no pun intended, of a game where his favorite team, the one whose insignia is emblazoned on the front of the dorky hat he insisted you wear, is tied with the dread rival team. When the blurry vision that usually accompanies my migraine coincided exactly with the twenty-minute laser light show at the end of the second period, I did what any self-respecting person would do in my position. I took eight more Maximum Strength Multi-Symptom Midol and kept my mouth shut.

I arrived home at approximately 11:00 P.M. I tiptoed about, ate a Reese's Piece, and proceeded to bed. I turned on 1010 WINS news radio because two minutes of it and I am usually out like a light.

So there I am, waiting to drop off, when suddenly I realize I've heard the traffic and weather twelve times, the news eight times, and the sports twice, and although I am lying stock-still, it is with all the serenity of a stunned deer on I-95. I begin to sing silently to myself in the hope that a gentle, regular rhythm will lull me, but find that I am only able to produce rousing renditions of "Those Were the Days," "Hava Nagila," and "Onward Christian Soldiers," the singing of which convinces me that they are, in fact, all the same song.

Just when you'd think I should give up all hope of ever falling asleep, lo and behold at 3:34 A.M. I do. And in what seems like no time, because it was no time, I open my eyes and it is 4:19.

At approximately 4:53 A.M. I cry, this lasting only a few seconds due to the fact that Midol relieves bloating and I imagine myself bordering on the dangerously dehydrated. Five-fifteen finds me lying crosswise on the bed gasping dramatically for air and flailing my arms and legs, an unpleasant reminder of how my single status makes just that sort of self-indulgent behavior possible. Five-forty-seven and I am searching wistfully out the window for signs of life. There is, in fact, a lone light shining from across the alley, and I momentarily hallucinate I am in college pulling an all-nighter, which makes me even more agitated because I can't get the paper done in time and every day it is late I lose a grade.

At 6:02 I yawn. My first yawn of the night slash morning. You can imagine my disappointment when I recognize that this is not a promise-of-slumbers-to-come yawn but rather an I'm-bored-out-of-my-skull yawn. The kind of yawn that says: Fuck the warning on the package, I could definitely operate a forklift.

At 6:34 A.M. I turn on the light and reread *Our Bodies, Ourselves*.

At 7:00 Erica's alarm goes off. I open the door of my bedroom just in time to see her slap the sleep button and roll over.

At 7:08 her alarm goes off again and again she goes for the sleep button, but because I have moved it out of her reach she knocks over the lamp. At this point she gets up. I brief her on my previous night's activities and I eat a couple of bowls of Captain Crunch in order to compound my exhaustion with a sugar headache. We determine an investigation is in order and no sooner do we set our minds upon the task than we conclude the fiendish culprit is none other than Maximum Strength Multi-Symptom Midol. Ten caplets, to be precise. Close examination of the package reveals that in addition to the nonaspirin painkiller acetaminophen, each Midol caplet contains sixty milligrams of caffeine. With little or no pharmacological training we further deduce that the six hundred milligrams of caffeine I had ingested within a two hour period was perhaps enough to rouse a dead horse.

Why had I done it, I asked myself. What had I been thinking, besides the obvious things like how's my breath and did I shave my legs this morning just in case? In my mind's eye I retrace the fatal moments which had brought me irrevocably to this moment, this day. This day I thought I might never see. This day which was separated by forty-five minutes from the day before it.

I did it for a guy.

Phew, what a glaring, harsh moment of self-awareness this is. I wonder if there is any way I can tell this story and have it seem like I did it in the name of those dolphins that get caught in tuna nets. Or World Peace. Stop the fighting now, or I overmedicate!

At least I got the last laugh: the Rangers lost.

from the ashes like
the phoenix

———

I'M not sure why it is that by this time in my life I am not either a VJ on MTV or living in the Mexican rain forest with my artist husband and young son and doing some kind of socioenvironmental study or something. Where is my drive? For God's sake, why don't I go get my Ph.D.? When did I fail to take the bull by the horns and become a successful human being? I really mean it when I say that I'm not going to get through the rest of this day if someone doesn't call who is more of a complete failure than me to tell me how lucky I am that I'm me and not them.

Really, I just heard that someone I was friends with in high school is living with her husband and baby in the Mexican rain forest. What the hell are they doing down there? Who even knew Mexico had a rain forest? The last I heard they were just obscure, meandering artists. She was a dancer and I don't remember if he was a drummer or painter, but either way, when did they become sociologists? It's like me all of a sudden saying I'm a brain surgeon. It gives me the same sinking feeling I got when I was in college and there was a boy I liked and my roommate would accost him at a party and, in the name of friendship, extol my virtues in a manner that would make her breasts appear larger. How she managed this I'll never know.

My mother insists there is no such thing as success or failure; all there really is, is just living your life. I completely disagree. And even if it's true, no one in their right mind ever feels that way, so what's the point? No one I know is really happy, at least I hope they're not, so why pretend things are otherwise? Go ahead, try to walk around happy. Try to accept things as they are, enjoy your day, be thankful for your blessings. Go ahead, you'll just be riddled with guilt because you don't deserve it or you gave up on your real dream or because people are homeless.

And even if it turns out that this old high school friend of mine and her husband and her child just sort of wound up in Mexico, aimless, broke, by default, maybe it's a bad scene, whatever; still, I know they will return to civilization in due course and she will choreograph *The Scourge of the Rain Forest* and will win a Bessie and a Pulitzer and his paintings of her and their son, Leaf, or Rain, or Monkey Boy, naked, dancing in the Mexican wilds will become collectors' items. Their exile will have fed their art, and they will ultimately write, and it will also be written about them, that had it not been for those dark years, well, you know the rest.

It drives me crazy that you can't count on anyone to stay down so you can feel up.

And just to put a capper on the whole thing, tonight I have to go to a party. I've got to get in the shower and dry my hair and put on makeup and pick an outfit. And then for three or four excruciating hours I have to skirt the edge of my bitter homebody character and try not to get so smashed that I say something truly revealing. Because nobody wants to hear it.

I am currently a freelance recruiter for a well-known retail company. I do this between auditions and acting jobs because it pays really well, much better than waitressing, and I don't have to work at night or on weekends. Also, it uses a portion of my brain that I did not heretofore know existed and which I plan to shut down for good as soon as is fiscally possible. I know nothing about retail—I don't even like shopping—and yet it is my job to ingratiate myself with secretaries and assistants at rival companies until I have learned the names and titles and telephone extensions of all of their executives— information which is often considered proprietary. I have several nom de phones, and sometimes I say I am from the Association of Retailers/Retailers of America/American Association of Retailers and am sending reports/invitations/

holiday cards. Or if someone sounds coldy, I say, "You need those Kleenex with the lotion." And "Go home, for God's sake." We laugh, and I get the names I need. I then call the executives and we discuss opportunities with my company. I work anywhere from ten to thirty hours a week, depending on my workload and on my acting commitments and on my mood. There's only so much of it a person can take. *This* person can take. I know there are people, reasonable people, people with whom I work, who devote their lives to this. Crikey.

The recruiting job sees me through many financially unre-warding theatrical enterprises. And although I am getting what seems like a reasonable amount of work, I succeed in staying just beneath the radar of almost every agent and cast-ing director in the city.

I get a commercial agent, though, and book a commercial for an East Coast supermarket chain. The commercial shoots in Wilmington, Delaware—a delightful city—and I finish the first day of shooting trying to get the feeling back into my fingers after spending ten hours reaching into the glass-front cases of what is commonly referred to as Your Grocer's Freezer. I spend the second day of shooting jumping on and off a rolling metal cake rack as it careens through the bakery section, a low-rent version of the old Rice-A-Roni ad. I do this dressed in a skirt and pumps—I am a working woman—and as I roll along I gape at a selection of pastries and sweets and compliment the store's lighting. Thankfully, the commercial doesn't air in New York, although my Philadelphia relatives are *very* impressed.

I do a commercial for ESPN in which I stalk an old, odd-looking sportscaster. It is a very funny ad and I hope it will do for me what that Dentyne commercial did for the actor Rob Morrow. It doesn't. The only people who see it are the type of sports fans who watch ESPN in lieu of a career.

I get a part in a really good off-Broadway play. I play the lead character's talking dog. This is not the same play in which Sarah Jessica Parker is transformed into the sexiest talking dog on earth and almost steals a woman's husband from her. In this play the old, mangy talking dog dies.

I play at least one lesbian a year in one or another downtown theater. Most of these plays are written by the same playwright and are attended by her large lesbian following and by my family, who surprises me by roundly accepting my work-related lesbianism. Their only request is that I provide them with a project-by-project breakdown of sexual content so they know which friends to invite. Some don't want to see any kissing.

One of these productions leads to an audition for the lead in an independent feature. I have already been in one independent film that actually made it to the Sundance Film Festival, but my part in it was so small that my character did not have a name. On my résumé, though, I call her Janice.

The character I am auditioning for, Sarah, is the out-of-work wife of a millworker and she is having an affair with

Susan, the college-bound granddaughter of Sarah's black neighbor. The focus of the story isn't the sex, the director assures me; it is about Sarah wanting to *be* Susan. Sort of a Catskills version of *Persona*. After the audition I go home and read the script. There are a series of funny, pathetic scenes of Sarah lying her way through job interviews in a crumbling, town-that-time-forgot town. I am interested, that is, until the scene where Sarah has sex with Susan and has to simulate orgasm-face on-screen. Is it a tasteful but explicit scene? Or is it soft girl-on-girl porn?

The director calls and tells me I'm the front-runner for the part of Sarah, and she asks if I will come read with a bunch of prospective Susans. The director and her crew are now ensconced in a room in the offices of a film company on Washington Street. The building has been renovated expensively using a lot of exposed cedar beams and moldings and it smells like a freshly cleaned hamster cage. I read scenes with a succession of attractive black actresses, the last of whom makes the greatest impression. She is beautiful. She wears boys' suede sneakers, jeans low on her straight hips, a tank top, and a baseball cap on backward. I could never pull that off. And her acting is very sullen, almost as if she isn't acting. As it turns out, she isn't. She's a poet and she has never acted in her life. This bit of information will eventually be the source of one of Life's Great Ironies. If you can't guess what that might be, don't worry, I'll fill you in later.

In late August, the poet and I are officially cast. Rehearsal is a strange affair. Nonactors can be more temperamental than actors. It's not their fault; they don't know the protocol, like the one where you act with the other person in the scene.

. We shoot the movie over the month of October in upstate New York, the cast and crew all living together in a big house. Someone shoots at us through a window one day because they think we are making pornography. My greatest success is that I manage to convince the director that the orgasm close-up should be of the poet's face, not mine.

We run out of money.

A year later, with more money, we return to the Catskills and finish the movie. In the hiatus the poet has taken acting classes and becomes more fun to work with. We get into a bunch of festivals, including a prestigious festival in New York. The film gets some very nice and some okay reviews. I get some very nice reviews, too. My friends call me and say, "This is it, your life is totally going to change."

In fact, nothing happens. Well, one thing. Two months after the film screens in New York, my father-in-law passes away. He was a fairly conservative, old-school gentleman, and it is intimated more than once that seeing me locked in an erotic embrace with a black woman hastened his decline.

I finally get a theatrical agent when I perform a piece of my own at a reasonably hip theater company and am a hit. There is a sense of legitimacy an actor gets from having an agent, and all these years I have wondered what that feels like. Now I know. It feels mostly like my agent sends me out for things I am not pretty enough for.

A year later the poet appears in a highly publicized film at Sundance. Her picture is in the *New York Times*.

Maybe the best I can hope for now is that everything will collapse and I will have to move, I don't know, somewhere horrible, so I can rise again from the ashes like the phoenix, and I, too, will write of those harrowing days and nights and, who knows, maybe also start a grassroots organization that in five years' time will make me both financially independent for life and a good person.

jack has a thermos

NOBODY loves a gadget like my father. Right up there with things that taste good are things that *do* things—unscrew tops, uncork bottles, inflate rafts, deflate rafts, juice oranges, blend milk shakes, vacuum spills, water lawns, keep tennis balls fresh. Like some kind of one-man *Consumer Reports*, he has tested the usefulness of several decades' worth of power tools, barbeques, and exercise equipment, the last in a never-ending quest to get in shape. I remember in the late seventies there was a rope thing that you attached to a doorknob and you put your foot in one end of the rope and your hand in the other and when you pulled with your arm your leg rose up and vice versa. It was simple, really. You just laid there on the floor, watching TV, using your own body parts for resistance. Unfortunately, the doors in our house had warped smaller from the air-conditioning and when you pulled on the rope you also pulled the door open.

Most people think of gadgets only as novelty items, clever contrivances that are possibly of more interest than of actual use and which make good jokey gifts at Christmas and Hanukkah. But as far as my father is concerned, any device, large or small, that promises to accomplish a task better than whatever device came before it qualifies as a gadget. Call him a technophile, call him a prophet, call him a sucker. I call him the Gadgeteer.

Over the years the Gadgeteer has bought untold numbers of flashlights and telephones and popcorn poppers. Inevitably, our family had multiple generations of things—five juicers in ten years representing biannual improvements in juicer technology. We had a camera for every occasion, a small sampling of which included a classic thirty-five-millimeter Nikon, the very first model of the Polaroid camera, purchased on a trip to Florida while they were still unavailable in New York, and a spy-sized Minox. On vacations my mother referred to herself as "Sandy Hold This" because my father would invariably insist that she carry his cameras and binoculars in her purse or slung over her shoulders, an upper-middle-class sherpa.

Surprisingly, our kitchen and workshop did not boast a full array of Ronco products. That is because you had to mail away for those. Nothing is worse for my father than waiting for something he has suddenly decided he has to have. Part of the pleasure he gets from making these purchases is the immediate gratification of bringing them home as soon as he has discovered them. The rest of the pleasure is derived from the gadget itself, which is also, by its nature, a source of immediate gratification. You do *this* and then *this* happens. If it takes more then ten minutes to accomplish a task, it is not a gadget. Also, if it is not fun to use, it is not a gadget. The washing machine and the dishwasher and the iron are not gadgets. The vacuum cleaner is not a gadget. Or rather, the *conventional* vacuum cleaner is not a gadget. The nifty, outdoor, suck-up-water-nails-and-industrial-waste vacuum cleaner *is* a gadget.

My father gets enormous satisfaction from the knowledge that he is taking care of his family by bringing home a better mousetrap. Many was the Saturday afternoon he returned from an excursion to the hardware store or the electronics store or a store in Westport called Silver's where pretty much everything they sold with the exception of luggage was a gadget, and presented us with his latest discovery.

"You've got to see this. It works like a dream. It's much better than that old thing we had. Wait a minute, let me just, wait, I think there are instructions somewhere, okay, this fits in here and this fits like this, there we go. Okay, now try it."

Sometimes the latest generation of a gadget had become

more complicated, enabling it to accomplish a new set of tasks and broadening the scope of its usefulness.

"Look at this, it has twenty different functions, twice as many as before. It's really amazing, the things you can do now."

And sometimes it had been simplified to an almost Luddite state of grace.

"Do you remember how the other one had all those buttons and dials and you had to set it every time? Christ, it was a real pain in the ass. This one you just turn on and off. You won't believe how easy it is."

Unfortunately, my father's technological prescience did not always result in a step forward for the Kaplan family. In addition to a museum-quality collection, Calculators Through the Ages, our basement housed a whole host of devices that were not so much functional as *hyper*-functional; they were too clever to actually work. Recreational items often fell into this category. Lounge chairs that doubled as floatation devices, safe versions of games that aren't fun if they are safe, science or craft projects that resulted in something completely unidentifiable. Occupying a prime corner of the dog's room—that was the room in our basement where the dog slept and where my brother and I were supposed to hang out with our friends but never did, opting instead for the paneled den with the big color TV—was a drawing contraption made of liter-size plastic bottles that you filled with water and hung from the ceiling and there were some magic markers and pulleys involved and some string and you were supposed to be able to draw enormous geometric shapes with the bottles and markers swinging in various directions. Sort of like a giant Spirograph. I am probably missing some important part of the description but I can tell you, I saw the thing in action and nobody *ever* knew what was supposed to happen. Then there was the giant chess set which consisted of a shaggy chessboard rug, about five feet square, and a set of ten-inch-tall, sand-filled chess pieces. They were hard and large enough that if someone threw one at your head it could make you cry. This was not a gadget in the literal sense, but was, from my father's point of view, a new way to play chess, and thus qualified as a gadget in spirit. I think he had a vision of my brother and me lying around on the floor, improving our strategic thinking while listening to the Archies. I don't remember ever playing a complete game of chess on that rug, possibly because at the time of its purchase my brother and I were totally consumed with making monsters by putting rubber globs in a heat chamber. Now *that* was a good gadget. The only member of our family ever to use the chess set was the dog. He lounged on the rug and occasionally carted the pawns from one end of the room to the other in his mouth.

My father's all-time biggest turkey took up residence in our garage in 1979. It was a Citroën with a hydraulic system that enabled the car to rise up when in use and lower down when at rest. The intention was, ostensibly, to give the rider the impression that he was floating gracefully above the roadway like a Thanksgiving Day Parade balloon. He wasn't. The Citroën was already so low to the ground that even levitated to its full height, you still had to crouch down to get in. It was a thankless car. It made my mother angry every time she saw it skulking menacingly on its haunches in the garage, and she was filled with dread at the thought of inheriting it when my father's muse, Gadgetella, struck next.

It struck all right but it didn't strike a car. Sometime around my sophomore year in high school my father decided that what we needed in our backyard was a hot tub. He and my mother had long debated the pros and cons of installing a pool, and one day, on the umpteenth fact-finding mission to the pool store, my father fell in love with a redwood hot tub. He went home and made drawings and calculated figures and he came up with the idea of having an eleven-by-thirteen-foot tub with a swim jet, the swimming equivalent of a tread-mill. He found a carpenter from the Old World who hand-cut and -fitted every redwood plank and built beautiful curving railings. In the first week after it was finished my brother and I stood at my parents' bedroom window and watched our father's white body flailing away against the foamy jet. That's about as long as that exercise regimen lasted. In all fairness, though, the hot tub was pretty great. If I'd been popular it would have been *really* great. And being Jewish, no one ever took their bathing suit off. Still, friends came over often and we steeped late into the night, playing with the jets and breathing in the bromine-scented steam. In the summer we turned off the heater and used the tub to cool off in, like a little swimming hole.

There were ways in which my father applied his gadgeteering skills with more serious intent. He was one of the first people to have a phone in his car. This was the late 1960s, decades before cellular technology, and it was essentially like having a shortwave radio in your car. He loved the idea that he could talk to a client or call my mother to say he'd be late without stopping to find a pay phone. The Mobile Phone, as it was called, looked just like the phones in our house, except it sat between the front seats in a box like the Batphone and to get a line out you actually had to talk to the Mobile Operator. At stoplights my brother and I used to open the window, hold out the receiver to the driver of the car next to us, say, "It's for you," and collapse in hysterics. In those days my father was a hero because he could stop for distressed motorists and let them call for help, and they often couldn't believe their eyes when he offered them the phone.

He was also on top of the computer thing very early, and both he and his business, a consulting firm, were transformed by it. One room of his company's offices entirely devoted to a throng of huge Burroughs computers, whirring and ticking and hacking away, floor to ceiling, twenty-four hours a day. They were straight out of a 1950s sci-fi movie, and it was not hard to imagine them eventually breaking free from their moorings and galumphing down the hallway to my father's office. He would yell "Aaargh!" and that would be that. "Some day," my father told my brother, who was something of a techie himself—he worked in the AV Lab in school and was one of the kids who wheeled the overhead projector into your classroom so the teacher could show you transparancies of Europe Before the Great War—"Some-day," my father predicted, "everyone will have a computer on his desk."

Now that he *has* one on his desk, it is continually breaking down because he refuses to leave it alone to do its job. There is always one more thing he can get it to do. "I was just trying to make the cross-referencing easier and I completely screwed the thing up. I was on the phone with the computer guy for half the day. But I know what I did wrong, so I just . . . come," he says to whoever is nearby, "I'll show you what I did." He just found some software that helps return the computer to where it was before he messed with it. "It's really fantastic. I can't believe it took them so long to come up with something like this. Before, you made one little mistake and you wanted to tear your hair out. Christ."

But nothing, no computer, no telephone, no waterproof wristwatch or rechargeable battery, no bagel slicer or egg coddler or collapsible beach chair with cup holder and shoulder strap, no fold-into-a-tiny-pouch travel bag, *nothing*, has captured my father's imagination to the extent that the thermos has. Is it too big? Is it too small? Does it attach to the car? Does it have its own carrying case? Does it keep things icy cold or boiling hot? These are the questions that consume him on a daily basis. And of course the thermos he is using at present is always the best one there is.

It is helpful to view it as a sort of mathematical model: Jack has a thermos. Maybe you have a thermos, maybe you don't. Either way, Jack wants you to have *his* type of thermos, because according to Jack, his is *fantastic*. Yours, if you are not just reusing spring-water bottles—which, I might add, Jack thought a very good idea for several months—serves your needs. Jack's thermos does not serve your needs, or rather, you don't give a shit about the whole subject. Still, Jack buys you his type of thermos. Both because he wants you to have something fantastic and because he wants to prove to you that he is right. You use his thermos a few times. It's okay. When you concede as much, Jack tells you he has found a thermos he likes better.

This is the way with my father. He is of one mind about things, *his* mind. It is impossible to convince him otherwise, even with charts and graphs and surveys and polls. No device or idea works until he sees it works (or until he reads the directions), and then nothing *else* works. And this applies to every aspect of his life. If he likes a particular food, for example, it is not possible to make him understand that you don't: "Rye bread is delicious, you're crazy." Each time my mother served the horrible Liver Meal—that was the meal consisting of sautéed chicken livers, cooked spinach, and mashed potatoes—my father would lean over my plate and stir the spinach and mashed potatoes together, thus contaminating the mashed potatoes and essentially doubling the amount of spinach. "How can you not like this?" he would say. "It's delicious. You don't know what you're missing."

When I was about twenty-four years old, I had an operation of sorts to remove a congenital something or other that had taken up residence in my right lung. Before the procedure began, my father instructed me to drink eight full glasses of water. He was absolutely sure that he had heard somewhere, or read perhaps, that prior to procedures "such as yours," a patient must deluge him- or herself with fluids. I told him that no one had suggested this to me but he was adamant, the result being that I later found it necessary to pee mid-procedure in the presence of ten or so medical personnel, onto a folded towel, which was all I was offered, while I lay stark naked on a gurney. A nurse said to me "Eight glasses of water? What in God's name for?"

Ahh, but the trouble is, I can imagine *myself* saying to someone: "You have to drink eight glasses of water this morning before your procedure." And when they ask why, I would posit: "Because this is that kind of procedure." Maybe I read it somewhere, maybe someone I know has had a similar procedure, maybe I made it up. (I am often most convinced of the rightness of things I make up.) I am, in fact, the Gadgeteer's daughter, and I have inherited a variation of the Gadgeteer Gene. While I couldn't care less about all those clever little devices, per se, I have an almost unshakable confidence in my ability to think logically. And I am pretty much always right. And I don't stop giving my opinion until I am sure it has been absorbed if not actually ratified by all listeners, or else they leave the room, which sometimes happens. My father is more dismissive of opposing opinions. Or rather, he is sympathetic. He will shake his head at the sorry state of your antediluvian gadgets and might even buy you some new ones, but he will not lose sleep over it. I, however, am less *sympathetic* than rather dangerously *empathetic*. I feel what you feel. And if I feel that you are wrong, I will stay up the entire night until I have brought you around. Nobody can beat a dead horse like me, nobody. Except maybe my mother, but look at what she's up against.

If I sound impossible to live with, I'm not, but there's time.

If my *father* sounds impossible to live with, he is, but now, thankfully, that is exclusively my mother's problem. It helps that my father is also the most loving, most generous man I know, and if you keep at him long enough, hard enough, if you are sure of your facts and armed with some evidence, you might occasionally break through to him. When you do he will laugh, wonderfully, at his own wrongness and say, "Really? I can't believe it. No kidding. I was absolutely positive. Oh, well." Once in a while he even proves *himself* wrong.

"It fit into this holder in the car door and would have been great except every time I took a sip, the goddamn thing leaked down the side. Christ. I got this other one, though, and it really keeps the water ice cold."

is that what
you're wearing?

———

I CRY intermittently in the car on the way to Greenwich Hospital to see my father, who has had a heart attack. I feel as though my head will explode from the surely combustible combination of a) relief that cigarette-smoking, bacon-munching Jack Kaplan has finally arrived at the moment he'd been gunning for all these years and was probably going to live to tell the tale and b) fear that the fact that I'm wearing blue jeans and clunky shoes to the hospital will aggravate my mother.

A typical telephone conversation with my father goes like this:

"Hi," I say. "How are you feeling? Do you feel better? Have you been to the doctor? Are you taking your medication? Did you read the thing in the *Times* about the reflux thing?"

"Where?" he asks. "I don't know. Did I? What day was it?"

"Yesterday in Health. Are you getting your esophagus checked regularly?"

"Yes, of course. How are you?"

"By whom? Doctor Yes-You're-Fine or a real doctor?"

"Aw, come on. By the Mount Sinai guy. I go every year. What's new?"

"You have to be careful when you have a sore throat."

"Any time I have a sore throat I take antibiotics."

"Antibiotics don't cure cancer."

"I guess you're right. How's David?"

"You've really got to lose some weight."

"I'm trying."

"Clearly you don't care if you live to see my children."

"Jesus Christ, how can you say that?"

"You'd stop smoking if you did. Anyway, when are you guys coming to see my play?"

My father is a notorious self-medicator and self-un-medicator. He pops untold numbers of antacids for his hiatus hernia and occasionally likes to go cold turkey on his steroid medication. Dangerous? You betcha. Once he decided to stop taking his iron pills (Why? Just because!) which caused him to lose a lot of weight very quickly, which made him happy even though he looked like death warmed over and, in fact, might have died. He is also fond of self-diagnosis. He likes to get his blood work sent directly to him by e-mail from the lab. Oh, look, his good cholesterol is up and his bad cholesterol is down. Oh, wait. Which is the . . . ? No, that's right. He is going to be fine! Between us we have raised Nagging and Deflecting to a high art.

The same character traits that make my father a wonderful father—his generosity and spontaneity, his utter confidence—make him a medical nightmare. He is at once self-indulgent and hypercritical. He is a dreamer with no patience. He cannot turn *except* on a dime.

I do a bunch of crying in the hospital parking lot to try to get it out of my system. My father is a bigger crybaby than I am (he teared up when he heard I had lost my virginity, possibly out of relief that I finally had) and if I set him off I'll start again, too. I am sure that he is already going berserk blaming himself for being in this mess. My father handles serious troubles by taking complete, furious responsibility for them. He handles the lesser ones by throwing responsibility onto my mother. Sometimes his recriminations will be so baseless that all he can spit out is a "Sandy!" in a disgusted tone of voice. Or the ever popular "Sandy, how could you let me do that?"

As predicted, my father's expression when my boyfriend David and I arrive is self-reproaching and slightly sheepish. He thought that since his father died at ninety-six he could lard up all he wanted. He could smoke and sleep badly and fret silently over work issues because he had good genes. Well, my grandfather did exercises every morning before breakfast and my grandmother never met a bland food she didn't like. So, my dad is both mad and embarrassed. This works well for him because he has a great face. His mouth and his brow will do one thing but his eyes another. He can also be both serious and amused at the same time, like Ed Asner. The first words out of his mouth are something like "Jesus, I'm such an idiot." Followed by "Whatever you do, don't tell my mother. She'll make us all crazy." He also feels terrible for bringing my mother home from Florida, where for the last two weeks she has sat vigil with *her* mother, who lies dying of cancer in a hospice. Yet he can't wait for her to come. When she does David and I leave so they can be alone. It is going to be a ten-Kleenex reunion and in the back of my little, little mind, I do not want her to ask me why I can't carry a real hanky. My mother isn't a shallow person by any means, but still, seeing my dad alive before she sees me may afford me a short grace period.

It's not really her fault; she hails from a long line of criticizers. Her mother was so disparaging of her hair that she became known in her family circle as Sandra *Your Hair!* Siegel. In my mother's adulthood, my grandmother likened her to Edith Piaf—which was not a compliment, since Edith wore all black and usually looked like hell. Despite the fact that I have memories of my mother, who is, in fact, gorgeous, in vivid Pucci ensembles, her long brown hair flipped fashionably up at the shoulders, overall, her fashion sense could best be described as Quiet Chic. Simple dark suits and elegant white blouses and little black dresses. My grandmother had short hair dyed various shades of sometimes startling red and she dressed very stylishly. Even in her old age she wore straw hats set at a rakish angle and high-heeled mules and colorful little ponchos she had crocheted herself. She never made her peace with the fact that my mother wouldn't take her advice. Once, my grandmother told her: "If you buy that rug I'll never be able to come to your house." My mother bought it. The salesman clapped.

In a bootless effort to rewrite her genetic code, my mother has managed only to reorganize it. Her technique could best be described as rhetorical-aggressive. She'll ask me: "Is that what you're wearing?" Technically, I am already dressed in the offending outfit. So is it possible she really wants an answer, as if it might be anything but "Yes"? As if I travel to various events with a Saks Fifth Avenue hanging bag of alternative outfits? As if I'm just wearing what I'm wearing to test her patience and embarrass her in public? At least my grandmother was direct. My mother just can't bring herself to say "You look like Patty Hearst." And my father has begun to parrot her, though he lacks her subtlety of tone. "What is that? Is that a skirt? Is that how that's supposed to look? Hah, hah, hah."

At the same time neither of my parents could be considered either a trendsetter or a clotheshorse. They advocate quality over quantity, so, for example, if a little black Valentino cocktail dress or a pair of Gucci loafers or that maroon cashmere V-neck can be respectably worn anytime in the current quarter century then they—that is, my parents, the dress, the loafers, and the V-neck—have all done their job. My mother buys the same three pairs of beige-y sling backs every few years, as heel styles change, at Lady Continental. One pair has a black calf toe, one has a black patent leather toe, and one has a black silk toe. Silk toes are the dressiest.

When I was young, we dressed up to travel. I wore loafers and kilts (Dress Campbell or Black Watch—I had both) and button-down shirts and sometimes a navy blazer every time we flew to Florida to see my grandparents or went on a vacation. My brother wore the same get-up minus the kilt but plus gray flannels and a tie. You'd think my mother ran a restricted private school. When we arrived in Ft. Lauderdale, we would invariably be greeted outside the air-conditioned Delta terminal by a blast of tropical air, and during the half-hour ride to my grandparents' apartment we would itch and sweat as heat billowed in through the car windows. Or worse, it would gust across the dusty tarmac as we sat out the hours-long layover in an airless departure gate in San Juan waiting to board a puddle jumper for some island or other. And, of course, we would leave the same way, waiting at the airport in a near-faint among families comfortably decked out in unstylish shorts and T-shirts and souvenir straw porkpie hats.

Once during high school I went shopping with my friend Sue, who was a dancer, and we each bought slinky Danskin-type leotards and wrap skirts at the only groovy boutique in town. The leotard I bought was a glossy brown and the skirt a glimmering peach with a ruffle at the bottom. I brought them home and showed them to my mother and they never again saw the light of day. I remember she sent me off to my ninth-grade formal in her full-length kilt. I'm sure it was very smart in the seventies for a grown woman to wear a little sweater and a full length kilt to a holiday party, but you can bet that no teenage guy was going to put his hands on my scratchy wool ass during "Free Bird," providing he could even find it under there.

In all other ways my mother is a model mother—loving, smart, supportive, fun but not so much fun that my friends prefer her to me. Doesn't she deserve every consideration at this moment? Now more than ever, with her dying mother and her seriously ill husband? Unfortunately, it is too late to go back and put on a kilt and loafers.

David and I take seats in the ICU waiting area. We talk about how nice Greenwich Hospital is. It is very calm; nobody is running around with the paddle cart yelling "Clear!," which confirms my theory that people in Greenwich die slowly and quietly of alcohol-related illnesses. My brother shows up with his wife and daughter and we hug each other in that awkward but comforting way known only to siblings. Then we all talk about how lucky Dad is to have had the good kind of heart attack, the warning kind, as opposed to the bad kind where suddenly your arm hurts and then you drop dead.

When my brother and I were growing up, if the phone rang after ten o'clock it could only mean one thing: the Grim Reaper had stopped in for a milk shake with the Kaplan family. Sometime after college (where your phone *only* rings after ten o'clock) this practice resumed. So, for example, the night my mother left an ultra-casual message on my answering machine in my Upper West Side studio apartment, time stamped 11:53 P.M., I knew my grandfather had died. Every family has its own encoded language. This was ours. "Hi, honey. It's Mom. Give us a call when you get home. We'll be up." Sure, they'll be up. They'll be up planning the funeral. Last night there was a similar message from Florida: "Mom . . . call . . . anytime." Tappity tap tap. I assumed my grandmother had died and started to cry. Boy, I loved her. Then the phone rang. This was unexpected. It was one-thirty in the morning. Grandma was already dead, or so I thought; what's the deal? It was my brother: Dad . . . heart attack . . . hospital.

After a while my mother opens the door and we go back into the hospital room. If she notices my Doc Martens she doesn't mention it. Actually, Sandy, herself, doesn't look half bad. She looks like she only *had* a nightmare as opposed to is *living* one. She is wearing her uniform: a straight skirt and a cashmere sweater with a silk scarf at the neck, pumps and stockings, simple gold earrings, and her travel talisman, her grandfather's gold pocket watch on a long chain around her neck. Still, something is not quite right. Shouldn't she be more of a wreck? Then I realize with a shock: *These are the only kinds of clothes she has.* But also, my mother doesn't like her children to see that she is worried. She got this from her father, who was never sick a day in his life, until he was and then he died, but you never saw him sweat. If you ask my mother how she is, she will invariably tell you that she is Fine. If she ever, God forbid, lay injured and starving on a mountaintop, and you happened to get her on the phone, she would still be Fine. I compliment her on her composure and she tells me that after hearing her story, the man sitting next to her on the plane told her, not unsympathetically, that she looked like hell and should go put on some makeup.

A woman with very, very long hair, the kind you can sit on, the kind your sister gets you on *Oprah* for so they can cut it off on national TV, arrives to discuss my father's diet. Now, anyone who knows my father knows that a discussion of his diet can only lead to misery. His favorite foods are Hellmann's mayonnaise and butter.

"No orange juice, no coffee," says the nutritionist. No smoking. (No shit.) No nuts, no ice cream, nothing fried. No butter or mayo or full-fat anything. Low- or no-fat everything. Chicken or fish cooked in olive oil. Or Pam. My mother, a purist who would as soon be caught using a cooking spray as she would wearing a wool/poly blend, nods appreciatively and mouths Yes. *Yes.* My father has his "I'm listening" look on. His brow is furrowed and his head is turned slightly, presenting his ear to the speaker. It is the same as his "I'm not listening" look. We all know his brain shut down the minute he heard "No." My father is never going to become a good heart-attack patient. He simply will not change. No matter how we will nag him and lecture him, bully him, flog him. Year after year, my mother, my brother, and I will continue the torturous dialogue.

"Dad/Jack, no!"

"What, I'm just having a bite."

"You're not!"

"Come on. I never eat this. I swear to God."

"Dad/Jack stop! Give me that."

"Jesus, you're a pain in the neck."

That night I lie in bed with my mother in the apartment in Greenwich. I wear one of her old flannel nightgowns, she wears another. It is strange. We can't believe we are in this predicament. We can't believe we are in Greenwich. My parents sold our house in Weston and moved here a couple of years ago, its greater proximity to Manhattan a concession to my mother, who hadn't been all that excited to move to Connecticut in the first place and who has probably dreamt every night for the past thirty years of owning an apartment on Madison Avenue. I am more like her than I am willing to admit. I'd like an apartment on Madison Avenue, too. I tell her I'm sorry about my ratty clothes but that I'd dressed that morning in a daze. She says: "That's okay. Who gives a shit?"

Come again?

One of the hardest things about growing up is how one day it suddenly dawns on you that your parents are human. It hadn't occurred to you before. Why should it have? But then something happens, some *thing* happens, and the veil drops. It may have been totally insignificant, like the way your mother ran her finger around the lip of her wineglass at dinner parties as if she were one of those water-glass musicians, or how your father mixed Bosco straight into the milk carton and didn't tell anyone else. Or it could have been something huge, like the night not long ago when your mother told you she felt like her dad was the only person who ever loved her exactly as she was. Or the moment you realized your father was truly *incapable* of changing, even to save his own life. These are just moments, really, blips on the parental screen, during which they reveal their humanity, and that they are in the world, flailing about as helplessly as everyone else, everyone who is *not* your parents. Blowing it. Surviving. Hanging on by their nails. That they are at once more spectacularly resourceful and more deeply flawed than you might have ever imagined inspires both scorn and admiration, two emotions you'd always reserved for nonrelatives. But, happily, between the blips, they are just the same as they have always been, annoying, yet impeccably dressed, and you breathe a sigh of relief. It is too painful for them to be human.

On the rare occasion when my mother looks a little unkempt, I become extremely nervous and start calling her twice a day, and whenever my father suddenly loses weight, I suspect something shifty is going on and grill him so mercilessly that he asks me if I have any plans to leave the country. But you'd think the two of them would have realized by now that I'm *supposed* to be a mess; I've always been like this. They can't still be surprised. If I clean myself up, *that's* when they should worry. At the center of the whirling vortex of our mutual disapproval is the notion that if we all *look* okay, we all *are* okay. What a dilemma.

Not for my future mother-in-law, though. One night, a little over two years after all of this, shortly after her own husband died, and on the verge of a sleeping-pill-induced sleep, she said to me, "I love you, but I hate how you dress."

they weren't brave

———

IT is raining when our plane touches down. I look through the double airplane window, wondering how bugs get trapped between the two panes, and I see that it is raining, and I know I've made a colossal mistake, an epic mistake. There is a strange man sitting beside me and he is holding my hand. I have no idea who this man is; what is he trying to do? *Comfort* me? And although I can clearly recall that yesterday, following a brief proclamation from a rabbi, I had agreed to spend the rest of my life with this hand holder, that is hardly reason to trust him with my life.

I am sure every woman at one moment or another tells some version of this—the husband-cum-stranger bit. It's a bonding ritual between us, as satisfyingly anecdotal as the locker-room brag. Perhaps more so: I didn't just fuck him, I *married* him. But then, oh shit. Newlywed reality is a plane ride to a foreign land and the person sitting next to you is not your mother.

The first night of my honeymoon is spent in San Jose, the utterly charmless capital of Costa Rica, in a bizarre, slightly squalid hotel designed to look like a bunch of beach shacks surrounding a tropical courtyard. The sliding glass doors of our room face, at ground level, directly into this little commons, which has a pebbly sort of floor and is planted with an array of exotic-looking plants and flowers in order to give one the impression that one is outside. One is not. There is, however, a thatch-roofed bar in the center of it all, where one may avail oneself of whatever tropical drink will best make one's immediate environs disappear. I make it clear to David that a) I may never forgive him for going to China on business and leaving me to arrange our honeymoon and b) we won't be having sex here. At bedtime I send him to the *conciergo*, or whatever, for new sheets because the ones on our bed are nubby. I sleep as if something hanging over my head were thinking about dropping down.

The adventure begins!

The big plan is to drive around the Costa Rican countryside, stopping first in the mountains, then heading southeast to the beach, and finally capping the whole thing off with a two-day white-water rafting trip through the rain forest.

According to our map, which is about as detailed as a paper place mat in a New England seafood shack (how else would one know where to buy the best saltwater taffy?), there are only five roads and ten towns in all of Costa Rica. The only way to get anywhere is to ask directions, but this is something of a trick question for locals. You can't give them your final destination and expect to get there. You'll end up driving for days and if you're not careful wind up in Nicaragua. You must proceed village by village, stopping in each to inquire of the next. This is how I learn the only Spanish I know besides *huevos rancheros* and *Arriba! Arriba!*: *Qual es el camino a . . . ?* My new husband, however, speaks *uno petito* of Spanish and we slowly make our way north without running into any contras.

The roads themselves are terrible—very narrow and full of crater-size holes. Costa Ricans are resistant to the ever-encroaching tide of tourism, so there is probably someone in every village whose job it is to "maintain" the roads with a shovel and a pick, and he is reprimanded if fewer than thirty rental cars per week break an axle. And nearly every turn is a hairpin. I don't know if it is the fever caused by the bronchial infection I developed two days before our wedding or the nausea I feel as we navigate the serpentine mountain roads in our Jeep 4×4, but every time I look over at my husband I feel sick.

Who the hell is he, anyway? I am consumed not with doubt, exactly, but with self-consciousness, which, I think guiltily, is the last thing I am supposed to be. I am supposed to be totally comfortable and at ease and having the unselfconscious time of my life. We're finally alone! Tra la la. And it is either look at David or look a foot or so to my right, where the side of the road meets the sky. There are no guardrails and if for some reason we were to be forced off the road, say by a careening livestock truck or an infrequently serviced tour bus, we would end up two thousand feet below in a ravine.

We wouldn't be the first whose honeymoon ended in some kind of obscene tragedy. I've read of brides and grooms held up by gun-toting *banditos* or entwined for all eternity in tangled car wrecks. Why shouldn't we sail silently off some cliff? What will we have left behind? Certainly not each other. That's good. And our families? They'll probably be relieved. No one really wants to be an in-law. And our hopes, such as they are, will never be dashed, nor our ambitions thwarted, nor our careers come to nothing. We'll never get resentful or indifferent. We'll never divorce each other and marry a younger woman and an older man. What a load off.

Am I the only one for whom the most mundane of activities lead inexorably to accidental death? Of course, it would certainly be a shame to die on the way to an adventure travel experience. Ignominious, in fact. *They weren't brave; they weren't even there yet.*

Our few days in the mountains are primarily spent reading, eating, and sleeping to the thrum of a continual tropical rainstorm. I am secretly relieved that the potential for a mud-slide has preempted a four-mile hike up the side of a water-fall, and that the active volcano the Arenal region is known for is completely socked in by clouds. We drive toward the gray mass anyway, and settle for a few minutes of wistful gazing from a low-lying field where I pretend to be disappointed not to see molten lava shooting sparks into the sky. Had the night been clear we would have driven up a narrow, winding road, littered with tree branches and rocks, to have dinner at a restaurant with a world-class view of the fireworks. Ah, well.

When the sun finally comes out we wend our way back down out of the mountains, and after a blisteringly hot five hours on a sceneless (is that the opposite of scenic?) highway, we arrive in the coastal village of Quepos. My fever is gone and I'm feeling okay about David. We exchange some money at the local bank, which is guarded by two men with semiautomatics, and then buy some Mirinda orange sodas (the gustatory high point of our honeymoon) and some plantain chips from a little grocery. I want desperately to buy a box of Captain Crunch but it doesn't seem sporting. We walk around the village and when we don't find any cute shops we get back in the car and drive about a mile up a hill to our hotel, Le Mariposa. Although Le Mariposa is considered one of the best and most exclusive hotels in the region, breakfast here consists of hard-boiled eggs, fruit, dinner rolls, and Kraft-like slices of yellow and white cheese cut on the diagonal. There is no room service and extra towels are a hot commodity, but the rooms are spacious and exotic, if you consider the seventies an exotic decade. Every day at three o'clock hundreds of some sort of monkey fly through the trees outside our balcony on their way home for lunch. Despite plenty of signage asking them not to, the tourists often feed the monkeys, and sometimes, when you're on the beach at monkey-lunchtime, they drop out of the trees right onto your head.

On the the second night of our stay at Le Mariposa I am attacked by a ten-inch Costa Rican grasshopper. Apparently a fugitive from some nearby entomological freak show, it appears under a sconce on the opposite side of the dining terrace to the amazement and curiosity of all. Just as David assures me that it would not likely make its way to our dim corner, it promptly lands on my salmon en croute. I jump, nearly upending our table, and reel away with my hands over my face, like Tippi Hedren in *The Birds*. It would have been funny, had it been funny. A waiter comes over, captures the thing, and carries it off. Some time later, as David and I lounge romantically by the pool, it attacks my hair. I scream at the top of my lungs while executing a variety of graceless evasive maneuvers. Finally it lands in the blue water, and, because there is a God in heaven, drowns.

Why aren't I at a nice resort? Why am I in the middle of nowhere, eating freshly killed pet chicken sandwiches at "family restaurants" that consist of one picnic table surrounded by the squawking of the remaining pet chickens? Why am I in a country where I consider an Oh Henry! bar a suitable reward for surviving an afternoon of kayaking over giant swells in the Gulf of Mexico? Who am I trying to impress? David? Of course, David. I have known him for a year and a half, and I am still trying to convince him that I am of hardy stock. I *want* to be of hardy stock. I want to fit in with his friends, who are of hardy stock, who ski off-piste and hike up Mount Olympus or some such and camp out in places whose names in English mean "The End of the Earth."

Through the tropical nights I dream of a paddleless double kayak washing up on a deserted shoreline, a lone Teva floating in the waterlogged front cockpit. I dream of a vicious, fatal attack by starved orangutans. Of searing volcanic ash and washed-out mountain roads. Of kidnapping, carjacking, plane crashing. Of contaminated water. Of poisonous red ants.

Probably the only reason I can function at all on a daily basis is because my *body* has courage. It does things—drives, flies, *lives*—in spite of me, maybe to torture me. To get back at it I conduct psychological experiments on myself: I have written the obituaries of myself and everyone I love. I have contracted every conceivable fatal illness and been the victim of nearly every form of accident with the exception of an anvil falling on my head, which even for me seems farfetched. I don't do it to get attention. I do it to anticipate the grief, to see if I could live with the loss. I'm the worst kind of empathizer; I feel despair over unconfirmed future events.

The only time in my life when I wasn't fearful was when I was lonely. High school, most of college, my twenties. When I was lonely my loneliness was so big there wasn't much room left for illogical concerns like will the ancient gargoyle be knocked from its perch by the drunken frat boy and give me a fatal head injury. There was only room for: here I am, drunken frat boy. In my twenties, my loneliness acquired a flapping, manic, jokey quality that was in itself scary enough to keep other fears at bay. Then I met David and there was some joy. But joy is no match for worry, and the two of them duke it out in an endless refrain. Guess who wins? And, of course, the last thing I needed to worry about was a *husband*. Maybe he'll just quietly go away.

I suppose I agreed to go white-water rafting on my honeymoon because I'm an idiot and because I'd bragged about being an expert canoer. I am, actually, a certified canoeing instructor and I wanted this to mean something, to *count*, if anyone besides me is counting.

We depart San Jose for our launching point in the rain forest in the pouring rain. After an hour or so the bus stops at a large lunchroom where everyone but me has rice and beans and the guides await a phone call from the tour operator, Aventuras Naturales—that's Natural Adventures. There seems to be some concern about how high and fast the river may be as a result of all the rain, and about the lightning, but the word comes that we will press on. How this decision is arrived at I do not know. Maybe they figured that nothing the river could do to us would be worse than going back to San Jose without a hotel room. My feeling is that nothing the river could do to us would be worse than going back to San Jose *with* a hotel room.

It is still raining as we set off in our rubber boats, and the idea of a rain forest finally makes sense to me. The river is, indeed, fast and rough and almost immediately we are drenched with cold river water. In fact, there is so much water coming at us from all directions that it soon seems as if it isn't raining at all. And we're really flying down the river, but no one seems worried about it. We're all forward-paddling and back-paddling with enormous zeal, smiling and whooping each time we emerge intact from a hairy section. It's pretty exciting.

After a while we reach a quiet stretch, and all the rafts pull up to the shore. Some German hikers and their guides are just up the river taking turns riding a bodyboard type of thing through the rapids and David, who in his head is in constant competition with an ex-marine buddy of his named Bill, feels he has to try it or be forever branded a sissy. The guides throw out a safety line at the end of the run, to pull him in, but they miss him, and he continues to float down the river toward the next set of rapids. Just in time he gets caught in some branches and is able to hang on until the guides arrive. He loses a Teva, though, and it sails away without him into oblivion.

Dinner that evening is cooked by the guides in an open, two-level structure on the riverbank, and later they entertain us by throwing a cat off the second story so it whizzes, screeching, by us as we play hearts with another couple on the lower floor.

It rains through the night.

When I told my friend, Lynn, about the proposed whitewater-rafting portion of our trip, her only comment was: "You're a good swimmer, right?" But as we continue down the river on the second day and I hear the fierce, panicky scolding of our river guide—we have somehow managed to let ourselves be carried to the wrong bank of the raging, rain-swollen river despite his expert instructions and are now back-paddling furiously, desperately hugging the mossy rock wall that banks the river, poised at the edge of a swirling black hole—it occurs to me that the point is moot. There will be no swimming. The previous day's exhilaration has morphed into a horrible dread. The river is deadly; the guide is afraid. Christ, could there be anything more ill-boding than the fear of a professional? It cuts a swath through me. If the boat flips, if we are all tossed out, we will never recover in time for the next set of rapids. I feel a familiar heart pain in my chest and am for a millisecond actually comforted in the knowledge that it is induced by real, live events.

Earlier, our guide had jauntily disclosed to us that in the United States we would not be on a river like the Rio Pacuare. At the end of a heavy rainy season (end?) it offers class four and class five rapids—class five rapids are for professionals only—but liability lawsuits are not a fundamental part of the Costa Rican infrastructure, so here we are. Now, without a jot of jaunt, he warns, "We do not want to be on the far side of the river, you understand?" But then, funny, there we are. "We do not want to go under that waterfall!" What waterfall? I look up. The sharp peripheral spray of it stings my cheeks, and we float beneath it, unequivocal. The guide screams at us. We paddle literally for our lives. The front of the boat bends ninety degrees into the vortex; half of us are ejected into the roiling foam. The noise is immense. Rushing water is loud, you know? Niagara Falls must be deafening. What kind of honeymoon is that? It is the sound of death. Which leads me to the fact that here, finally, is the moment I have been waiting for. My body has caught up with me. Or I with it. Whatever.

It is much quieter down below the bubbles. I don't know where up is, but the guide had said, "Try not to panic; the river will spit you out." So I don't fight. I wait, patiently, in the swirl, to see what will happen. In time—it seems like a long time but I know it is not—I am spit out. David is too, but I wish I had seen him down there: *Here we are, together.*

* * *

Long afterward, after we were all back safely in the boat; after we paddled to a place where the river slowed and yawned and we rested beside a tiny piece of shore; after I sat on a rock with my head between my legs and my heart beating out of my chest like the heart of a lovesick cartoon character; after the raucous, celebratory dinner in San Jose with our new friends, the husband and wife who had snatched us from the calamitous froth (You saved our lives! We just grabbed and held on! Oh, my God! Cheers!); after we flew home; long after everything; I remembered the calm. I remembered the silence in my head. No endless, contrapuntal refrain. No fear, no loneliness. No false bravado. Peace. Release. Is that what joy is?

what happened after
the chicken crossed the road

———

For many years I believed that death would come to my New York grandmother, Dorothy Kaplan, in the form of a chicken. Maybe kosher, maybe not. I predicted that she would engage the chicken in a seemingly endless cycle of freezing and cooking until its molecular structure was so altered that it would combust spontaneously and consume the old woman in ball of fire.

I ate that chicken every time I had dinner at my grandparents' apartment, which, from the time I moved to New York after college, was about once a month, not counting holidays and other family gatherings. When I arrived in the late afternoon the table would be set and the challah would be growing hard in a napkin-lined basket. The chicken would have been in the oven since noon, defrosting and re-cooking, acquiring its signature teeth-gnashing consistency. On the plate it looked like petrified wood and in your mouth it felt like a ball of twine. That it had been basted with lemon juice and then lightly dusted with powdered garlic and paprika was not to its advantage. And how, after all that cooking, it would wind up on the plate at room temperature is still a mystery.

Over dinner, my grandparents and I would discuss books and current events, we would shake our collective fist at President Reagan, and, eventually, when we were done chewing, we would adjourn to the living room where my grandmother would play a little Beethoven or Bach or Chopin on the piano and I would dance in a faux-balletic style for the amusement of all. Finally, we would return once again to the kitchen for a slice or two of Grandma's special, still mostly frozen, dry-as-the-desert raisin cake, another victim of the infamous Cooking-Freezing Torture. In this case, though, the cold helped the taste. Actually, no food was ever served by my grandmother at a temperature others considered standard. But then, why should it have been?

Just because a person can't cook doesn't mean she's crazy.

When my grandfather passed away of presumably unrelated causes, the dinners increased in frequency and decreased, if that was even possible, in variety. But other meals were added, lunches at a nearby diner and Sunday brunches, which were rescued by David, who, to the endless delight of my grandmother, could make pancakes. Once in a while, praise God, we ordered Chinese food or pizza.

A little over a year ago, cashews appeared on the surface of the soil of my grandmother's house plants. Had she read somewhere that rubber plants crave salt? Perhaps. We did not question it. Who would want to know the answer to such a question? Next, the soil itself disappeared entirely from one large pot and the roots of the plant were wrapped in newspaper. Incredibly, it lived, unwatered, in such a fashion for perhaps eight months, maybe a year. Where did the dirt go? *Why* did the dirt go?

She stopped making chicken. She couldn't remember how. Cook it then freeze it? Freeze it then cook it? Does it need to be cooked at all? There was a book that everyone was supposed to read in high school called *I Heard the Owl Call My Name*. If you haven't read it here's the general premise: a guy hears an owl call his name and his days are numbered. I don't know if it really called his *name* or just hooted; I was too busy rereading *The Outsiders*. Anyway, one evening I was encouraged by my grandmother to sup on what I can only presume to have been a brand-new recipe for uncooked chicken. Note to fellow cooks: cinnamon is not an appropriate substitute for paprika.

A flare went up, a bell tolled. The owl hooted.

Not long after, I found my grandmother, a tireless, impeccable housekeeper, vacuuming the living-room carpet without the vacuum plugged in. We had what seemed to be an intelligent debate, which I lost, about the relative benefits of electricity. Happily, the ten-inch strips of carpet nap which ran hither and thither throughout the apartment and still comprise one of the sharpest visual images of my childhood—a strange green Cubist landscape—could be maintained by the pressure of the Electrolux alone, that is, sans Electro. Soon my grandmother couldn't tell a cabbie to take her to Lord & Taylor. She couldn't order French toast or blintzes or noodle soup. If she found her way to the diner without me she had to sort of *describe* what she wanted. Eyeglasses became teacups and telephones became combs and a lot of things became just "things." I was startled to notice one day that her *Times* crossword was filled in with nonsense words. She had taken her address book apart and only A through K and S remained. Thank God she could still play the piano, although the selection dwindled to four pieces, two Chopin, one Bach, and one rather dramatic-sounding Czerny speed exercise. They are all now burned, like the multiplication table, into my consciousness.

My parents came in every weekend from Connecticut and over time I found myself at my grandmother's apartment at least two or three times a week. She needed an escort around the neighborhood, a companion. I posted my phone number next to the kitchen phone, which I soon regretted. She took to calling me several times a day to ask when I was coming. She lost all track of time. Sometimes she woke up at three o'clock in the morning and thought it was three o'clock in the afternoon. She dressed and went downstairs and neither we nor the dark of night could convince her otherwise. If you told her you were coming at ten in the morning she would sit by the door for hours the night before, finally calling to ask in furious tones, "Where are you? Where are you?" Worst of all was that she knew something was wrong. Her head felt funny. Things were not as they should be, they were all off. She was becoming aware of gaps and was frustrated and angry and scared.

My grandmother's visits to the doctor now included a seemingly impromptu question-and-answer session between her and the doctor. The doctor asked: "Where do you live?" and "Can you draw me a picture of a horse?" My grandmother's answers were: "I live where I live" and "Are you crazy?" The doctor diagnosed her with Alzheimer's disease.

We had hoped that perhaps it was any number of other things: your garden-variety senility (whatever that is), a couple of small strokes or "episodes," or perhaps even depression. We wanted it to be something treatable. The only way to be absolutely sure it was Alzheimer's was a genetic test, which we decided against. She was very old. Let her go crazy in peace, we said. Let her go in peace.

And indeed she has, pretty much. For her. In her own

demanding way. A low dose of Zoloft helped considerably for a while, but the truth is she was never a peaceful person to begin with. My mother can certainly vouch for that. More weekends than she'd like to remember her in-laws parked themselves at our house in Connecticut. My mother would serve meal after meal, nosh after nosh. Over endless cups of Sanka my grandmother would quiz my mother on the price of everything in sight. What did she pay for the chopped liver? What did she spend on her hair? How much did the gardener cost? Wasn't that a lot? The only time she didn't ask was when my mother was giving her a cashmere sweater or a silk scarf or a Waterford bowl, although most of these things she put away and didn't use because they were "too good." Meanwhile, my dad would have come back from tennis or golf just in time for the afternoon nosh and a nap. Recently my grandmother has taken to referring to my mother as "the maid" or "the woman who works for Jack." The truth will out.

My grandfather, on the other hand, was a sweet, slightly hypochondriacal man who loved to sing bel canto songs (he breathed expertly from his diaphragm) and read whatever author was considered the hardest to read. Thucydides. Solzhenitsyn. He laughed at any joke that wasn't on him and occasionally some that were. He had large, smooth hands and a long, pale face with watery blue eyes and a majestic forehead. He resembled God without the beard. He swam and ice-skated until the age of ninety-two and when he died at almost ninety-seven he had all his marbles. The worst thing you could have said about him was that he told you the plot of the same opera every time you saw him and that he was a closet capitalist. He faithfully toed the socialist line until you put a porterhouse steak in front of him. The only person ever to do that was my father, who had his own business. My grandmother didn't buy expensive cuts of meat. And even if she had she would have cooked it until it was unrecognizable. Until it was a porterhut or a porterhovel. On one of the many occasions my grandfather held forth on the subject of his Russian childhood, he described his family as Gentlemen Farmers. "Ha," my grandmother spat. "Ha ha!" It wasn't a laugh; she actually said the word *Ha*.

My grandmother had her own brand of socialism. She really didn't like anything to be too good. Serviceable was the best. Nice was acceptable under certain conditions. Fancy was suspect. In a restaurant, anything more exotic than shrimp scampi was not to be trusted. And she was very particular, very judgmental about people. She was intimidated by education and wealth and often felt as if those who had them were lording it over her. She compensated for this with a sort of reverse snobbery. "She's a prima donna," she would say to you, under her breath, as soon as some unsuspecting acquaintance was out of earshot. Or some stranger; she didn't discriminate. Later, as her hearing began to fail and she refused to wear the expensive hearing aids I forced her to buy, this became problematic. The insults came loud and fast. "She thinks she's the queen." "La dee da." "How can a person be so fat?" She kept a leg up on people by withholding information from them, by not letting them in. Two days after my grandfather died, my grandmother and I ran into one of her neighbors. The woman kindly asked, "How is your husband?" My grandmother responded, "Not so well."

It has never been my belief that my grandmother and I were alike. I definitely *look* like her side of the family, but I have to hope that my personality descends from my mother's side. There is, however, a place where my grandmother and I come to a meeting of the minds. There is nothing more satisfying to her than when some fancy-pants is brought to justice. Like the voice-over says in *The Magnificent Ambersons*, people eventually get their comeuppance. Perhaps this was a holdover from her socialist beginnings, or perhaps it was a result of a life full of struggles. Whatever the source, it appears to be her legacy to me. I, too, have an innate sense of spite, an ongoing low-level resentment which is occasionally capped off with glee in the face of what I perceive to be the well-deserved ill fortune of select others. My grandmother and I are not vengeful, exactly; we prefer to wait for something more like karmic justice to be done. Besides, revenge is an act, you take it and you are done. Resentment is a state of mind, ongoing. Seething is something one can reasonably do for a lifetime, if one is so inclined. One day, many years ago, while we sat with my grandfather as he lay dying in St. Vincent's Hospital, my grandmother told me the story of Ceil, the Phillips' Milk of Magnesia heiress, who'd married into the family by stealing my grandfather's sister Sarah's boyfriend, who was the brother of their sister Beatrice's husband. (Don't try to parse that out; just move on.) The family agreed that Ceil was not a very nice person, but what really irked my grandmother was that Sarah's potential husband had been filched by a rich girl, a Capitalista. When Ceil's later life was marked by terrible tragedy, my grandmother was unsympathetic. "Ceil got hers" is what she said. Frankly, I'm usually satisfied if someone I don't like was reviewed badly in a play. I'm still young, though.

Despite all this, there were always people for my grandmother to have lunch or play bridge with, or attend a concert or visit a museum. Just because she criticized them behind their backs didn't mean she had no friends. In recent years, though, many of them have died or have disappeared into their own troubles or live too far away. And of course, my grandmother is further isolated by her diminished comprehension and impoverished vocabulary. I'd say her poor hearing and hard-to-please nature were obstacles as well, but they never stopped her before. Occasionally she tries to make new friends, but despite her best efforts things still go wrong. A woman in her building asked her to lunch but then only served her a glass of seltzer.

I tried to involve her with a senior group designed for the hearing and memory impaired. There were coffee and dry cookies, her favorites. There was mural painting and bead stringing and poetry reading. The people were nice. One woman spoke Yiddish to her. My grandmother smiled and nodded in response. While she was smiling and nodding she would turn to me and hiss, "Let's get out of here." She hated all these old deaf people, and she wasn't *game*. If she didn't understand the point of something, like stringing beads while you chat to make the time pass and keep your fingers nimble, she made a conscious choice to be too good for it. Also, I don't think she could understand or even hear a lot of what was said, which, of course, was the point of being there in the first place. It is too bad my grandmother never really understood irony. We only went to the senior group three times and always left within an hour. The last time was when I made her stay for the armchair exercises to music. Big mistake.

As my grandmother's dementia worsened, my family stubbornly contended that she still knew who we were. She might not remember our *names*, but she knew us. We recently came to the painful realization that often she has no *idea* who we are. My father is alternately "my son," "my husband," "my father," "the man," "the men," "Yashka," and in a spectacularly successful distillation: "The Jew." My mother is "her," "the woman," "Sylvia" (not her name), "Emma" (no), the aforementioned "maid," and, after a recent visit to Connecticut, "that wonderful woman" and conversely, "that bitchy woman." My brother, Steve, is "Bob."

I am "the girl," "one of the girls," "the tall one," "the short one," and "your sister," as in "I told your sister that I want her to have all the dishes." I nod in approval. That will be lovely, I say. Fortunately, because I happen to like those dishes, I don't have a sister. My grandmother doesn't remember that I am married, although she adores David. She cannot fathom the idea that we sleep together. Or that my mother sleeps with my father. (That makes more sense; she's the maid.) I asked her, "Didn't you sleep with Grandpa?" and her answer was: "No, of course not."

It has been our intention, through all of this, to honor my grandmother's wishes and respect her independence, but it has gotten to a point where she needs help. More help than any of us can give. Professional help. We considered hiring a nurse or attendant but she didn't want a stranger in the apartment. She wouldn't *let* a stranger in the apartment. Well, that is if you don't count the time, recently, when she asked a strange woman, a woman off the street, to come up and dust the furniture for ten dollars. The woman came. It's worrisome, I know.

My parents have found a good nursing home near them and have put her name on the waiting list. "Going to Connecticut" has become our euphemism for the thing we know she'll hate, but still we are trying to prepare her by talking it up as if it is something to look forward to.

"Wouldn't you like to go live in Connecticut?" I asked.

"What is it, in Connecticut?"

"It's very nice. It's in the country. Near Mom and Dad. There are lots of nice people. You'll make new friends."

"Oh, that sounds very good."

Thanks, perhaps, to the Alzheimer's, she is not suspicious of our vagueness. And I think, with her growing limitations, her own home has become a burden. Or something worse. She often says to me, as we sit in her apartment looking at the furniture and paintings, the books and dishes, the appliances and even the wallpaper, "I don't want it. Enough." And "It doesn't feel like mine."Sometimes she is more vehement: "Get me out of here, I can't stand it, I'm going crazy."

I am reminded of the plant wrapped in newspaper. Alive, or so it seems, despite the fact that the most elemental things are stripped away. Without the dirt and the water, without the *alchemy*, though, it cannot live for long. It is dying from the inside out and will eventually collapse. Implode. Is this the way of Alzheimer's? I don't know. It is just one way to describe the slow, dismal ruination of everything but the body, the shocking exposure of roots to air.

When she goes I'm not sure what I'll do. I have gotten used to the squeezing of everything into each day—work, marriage, Grandma. I've gotten used to the terrible dinners and the wonderful Chopin. I've gotten used to repeating myself, loudly, ad nauseam, and pretending to understand what she is saying when I don't have any idea at all. And I have gotten used to saying good-bye to her as she stands in her doorway, turning and waving every few steps until I am out of sight, as if to say I'm still here, I'm still here.

better safer warmer

It happened very suddenly and was, perhaps, the most violent act in which I have ever participated. A space became available and we had to grab it or risk waiting maybe six or eight months for another. I don't know why I thought there would be more warning. It's not like they call you up and say we're expecting Mrs. Feingold to sign off in about two months so you should start to get ready. Mrs. Feingold or whomever just dies and they clean up her room and they call you. That's it. So my parents and I packed up my grandmother's winter clothing, a few pictures and books and some of her music, and tore her out of the ground like a mandrake root, and although she came willingly the silent scream was there, not so silent; I could hear it, we all could. Despite our best efforts to prepare her, there was no way she could have known what was happening. And despite our best intentions, here is what *was* happening: Come, put on your coat, get in the car, you will never see this place, your home, again. Sorry, sorry, sorry.

At first she was excited. She was finally out of the apartment, out and on her way to, where? Well, to the good new place, we said. To the place near Mom and Dad. With a piano. None of us could say nursing home. But then after half an hour in the car she stopped wanting information and started wanting soup. When we got to The Place everyone gave my grandmother the big welcome. They gave her soup. They gave us all soup. We walked her around the common rooms and garden, which were beautiful. Nurses and aides and administrators made a fuss over her, which she loves. She was bubbly and charming and spoke nonsense and everyone thought she was adorable. My parents and I made furtive eye contact; it was going so well, but still, all I could think was: Just wait.

We went to her room, or rather her *half* room. It was small and the wallpaper was a psychedelic flower pattern that it seemed to me could itself induce dementia. The furniture was old and the bed creaky and hospital-like. I went out into the hall and cried. There was a roommate who was incensed to find people in her room. We drew the curtains that divided the space. At least my grandmother had the window half, looking out onto the garden. At least, at least. She had not shared a room with anyone besides her husband in seventy years. Now she was going to share it with a mean stranger. I went into the hall again and cried. My grandmother was taken off for evaluations of some kind. We waited in her room in a state of what can only be described as wistful dread. Surely the other shoe was going to drop.

It did. My grandmother returned distraught, furious. She hates to be poked and prodded and questioned and she was ready to go home. She said: "Get me out of here. You have to take me out of here. I've had enough. You can just kill me if you want but I won't stay here. I'll just lay down on the floor. That's all. I'll just lay down on the floor and die."

Now we had to explain to her why she wasn't going home, why she had to stay tonight, and maybe for a while. We didn't mention forever; what a ridiculous, damaging concept *that* is. My father left the room for a few minutes, maybe to figure out what would be the best thing to say, maybe to cry. While he was gone the roommate poked her head through the curtain and said, "It's a nice place, really." But my grandmother was inconsolable. My father came back and said quietly, "Mom, you have to stay." That was it for me and I went back out into the hall.

Somehow, a few weeks went by. At least one of us has seen her every day, even though she doesn't always know our names and will probably not remember that we were there. We walk with her in the garden. We sit and people-watch. We talk her into playing the piano in the lobby. (She complains that it is not as nice as hers, and she's right.) We have endless conversations with her aide and nurse and social worker so they'll know what she likes and doesn't like, who she is, so they'll understand her. My brother has come with his children, and even my mother-in-law, way above and beyond the call of duty, has visited.

She is still very confused by her surroundings, calls everyone names behind their backs (actually, she barely waits until they are out of view, much less out of earshot) and thinks the social worker is someone she knows from the beach. Sometimes when we arrive she seems all right, sometimes she collapses into our arms, lost and distraught. Each day she takes her clothes out of the drawers and the closet, folds them beautifully, packs what she can into the small straw bag I brought for her piano music, then leaves the rest piled on the bed and dresser. She carries her pocketbook everywhere and sometimes tries to go out the emergency door. She talks incessantly of going home (who wouldn't?), of getting on the train she sometimes hears in the distance and going back to her apartment.

This will change, we know. In time she will make friends and have activities. (Okay, she'll never have activities, she hates activities.) She will be cared for around the clock by professionals who are trained to meet the needs of people with Alzheimer's. We tell her: Stay for now, stay for the winter, it's better, safer, warmer, whatever. We tell her what we hope she'll understand. But she is planning a breakout, I think. *Dear Sarah,* (Who?) *Send help. I need help. Help please.* She has written this on a paper towel and put it in her pocketbook. She is our hostage. As I said, it feels like the most violent act in which I have ever participated. I would take it back, if I could.

hey!

———

H<small>EY</small>, you know how your doctor and your mother and all the books and everyone you ever trusted in your entire life and even your own common sense told you that when you got pregnant you would stop menstruating? They lied. You don't. You still get your period. It just comes out your nose. Do not be alarmed. You'll get used to it. About half the times you blow your nose, blood will pour from it in torrents. Some days big bloody clots will come out and on other days the blood will drip back into your sinuses and when you retch in the morning the blood will come back up and out your mouth and you will think that you are dying. Someone could make a lot of money in the nose-tampon business.

And you will have no air. No oxygen. Oxygen is carried

by the blood, and if you remember you don't have any of that. Every time you unplug your hemorrhaging nose you will look at the bloody Kleenex and think, there goes the day. There goes your healthful walk in the park, right into the trash. Some days you're too tired to speak. You just moan, which is an activity in itself. People who mourn the loss of their waists are just big fat crybabies. You have no bodily fluids. And yet you're puffing up anyway. You're like a big Rice Krispie. So why you're peeing thirty times a day really is anybody's guess. It's not like you're drinking so much water. (Eight glasses, eight glasses. Shut the fuck up.) Water tastes like pennies. It tastes like shit. A lot of things taste like shit. You eat so many saltines to try to keep yourself from vomiting that eventually they are what makes you vomit.

And you don't understand why, if the baby is using up everything else in your body, it can't suck some fat out of your thighs and ass. After years of vain secret gloating, suddenly the inner parts of your upper thighs are touching each other. And let's not even talk about your ass. Every day you are more and more convinced that that's where the baby is. Don't be at all surprised if one day your OB says that she's going to deliver the baby through your asshole.

And you keep reading about how by the end of pregnancy, this thing called the mucus plug (guess where *that* is; I'll give you a hint, it's not up your nose) pops out at some point to let the baby through the cervix. Unfortunately, you have not nor will you ever read how you spend your entire pregnancy making the mucus plug yourself by swallowing horrible, gaggy blobs of mucus that drip corrosively from your sinus, down the back of your throat, through your gorge, and if at that point you somehow manage not to vomit them up, they continue through your digestive tract and end up as the plug. (Don't quote me on this. It's only a theory.) And if you weren't exactly sure what your gorge was before there'd be no doubt about it now. At any given moment, when you're walking down the street, perhaps, or at lunch with your mother-in-law or auditioning to be the next Sprint PCS girl, it will rise up like Godzilla from the sea.

What else? There's more to come, of course, but why ruin the surprise? For now, let's just join hands in a circle and take a moment to curse the authors of that despicable tome—you know the one. The one that makes you feel like you've been arrested by the pregnancy police. The one that claims its special pregnancy diet is the solution to everything from nipple soreness to acid rain. The one that deserves to be remaindered into infinity. This is the book that purports to tell you what to expect. Let *me* tell you what to expect. Expect this five-hundred-page-long book about pregnancy to dedicate no more than two pages to nausea. Expect to be very very angry because believe me, ladies, if you haven't barfed up your own bile, you haven't barfed.

That said, the truth, the real truth, is that none of this stuff matters at all. When you walk down the street, if you can walk down the street, you will perhaps feel more special than you ever have in your life. Or maybe you won't. Maybe you've had sex with Mick Jagger and that brought you to the apex of specialness. Whatever. But you might surprise yourself. I love my growing belly beyond all reason. And I hate my fat ass a lot less than you would expect. Most days it is all I can do not to sit in the little chair and weep with relief and gratitude. Some days it is all I do. David, the Saint Bernard of husbands, will say matter-of-factly: "I can't comfort you in that chair. It's too small. You have to come over here." And he'll sit in the big armchair and somehow I will shlumph my way to him and sob in his arms. I will bury my soaking face in his neck and leave his soft skin glistening with my bloody mucus.

at the end of the day

———

IT is two-fifteen in the afternoon and my day just ended. Just like that, over. I was on the second floor of Bloomingdale's, where the inexpensive shoes are, waiting for the elevator to take me and my son—whom I had roped into accompanying me by virtue of the fact that he is five months old and doesn't know the phrase "count me out" yet—to the fourth floor, where the expensive shoes are, so I could buy a pair of hot pink shoes because I needed them to go with my new dress with the hot pink trim. When the elevator doors finally opened we could not get in because we had to wait for a man in a wheelchair to get out. It took a while. It seemed that he was blowing into a tube in order to operate the wheelchair.

He was pretty well set up, all things considered, or rather, no things considered, if you know what I mean. He had a little rearview mirror to look into and plenty of padded headgear to keep him steady as he looked and some tubes and then this breathing apparatus that appeared to tell the chair what to do, although when I told my husband about the chair he said he didn't think there was that kind of technology and I said I saw what I saw and if it wasn't his breath that steered him around than it was mental telepathy, jerk. But even if it was just a breathing tube (*just* a breathing tube, what is that?) and there was some other mechanism I did not know of, what did it matter when he had some children's clothing piled on his wheelchair tray? Because that is when my day ended. When I saw that he was shopping for a child. I said good-bye to my day, and my son and I got in the elevator.

Now, here's a question. Do we still go to 4 and look at pink shoes? The answer, of course, is yes. There is no point in having a dress with pink trim if you don't have any shoes to wear it with. Or rather, if you're lucky enough to have a dress with pink trim, it's okay to have the shoes, too. It's okay to be lucky. I think. Yes.

Maybe it depends on what your definition of luck is. You can't live everyone else's life. That would be terrible. You have to live your own. Everything is relative. Or everything *can't* be relative. That's it. Yes. Because either all of it is your fault or none of it is your fault. And it's not your fault. Well, maybe some of it is your fault. Angry, lucky, angry, lucky. You know what angry and lucky make together? Lungry. No, just kidding. Guilty.

You could blow your brains out like this.

Or you could focus. You could just think about your own situation and do the best you can and try to be grateful and productive.

You go all of your life wondering will this or that ever happen, will I be happy, will I get married or have a baby or even a car. Will I ever play tennis again? You ask yourself these things when you are alone and struggling and kind of poor. And maybe you worry about the tennis more than the others because it is less scary; it is the benign host of everything you always assumed you'd eventually have and be, that is until the day you get out of graduate school and take a job as a waitress to support your acting career, neither of which supports you or your tennis habit. The future is completely up for grabs. You date actors, who are notoriously both non-committal and broke, but they are sexy and they are there. It is a jovial but stunted culture. All that moving around and sleeping around and poverty. No one travels except for work. No one plays sports. No one has insurance. No one, including me, wants to do anything but act. There isn't room for the other things. Extraneous things. Frivolous, self-indulgent things. Like marriage and children.

And yet, here I am with my incredible son. What a fatty he

is. We spoke on that very subject just last night. I said, "What a fatty you are," and he said, "Aahroo" and "Mmnnn." He stroked my cheek and then he punched himself in the nose. We read *Go, Dog. Go!*

David and I started trying to get pregnant around the beginning of the 1998. I'd procrastinated—no, that's not it, what's another word, a word for being afraid you will disappear from the world as you know it when you have a baby even though the truth is that no one will even know you've been gone: woondangled or hublittered, something stupid and vain and meaningless. I assumed the first time we had unprotected sex I would get pregnant. I felt edgy and worried, like what if my big break, whatever that means at my age, nothing, comes and I am in the family way? Then of course my period arrived and I breathed a sigh of relief. When you are young you are taught that the easiest thing in the world to get is pregnant. So for a while it's the last thing you worry about. Or rather, you spend a lot of your twenties afraid that you are. It can happen at any time, even when you have your period. Even if you only did it once. You're a fertile field, a sea monkey, a packet of self-rising yeast, just sprinkle with water and watch things grow. But you're not, really, and of all the things that finally fall into place when you grow up, this is one that sometimes doesn't. The window of opportunity is really very small, you are informed, and getting smaller. Sometimes it takes years. Or doesn't take at all. There are statistics. You don't want to know. Suddenly, the worst thing in the world that could ever happen to you becomes the worst thing that could ever not.

The next thing I knew, seven months had gone by and every time I went into the bathroom and found blood I sobbed. There goes another baby, I thought. You'd think women would get used to the blood, wouldn't you? And we sort of do, most of the time, although we never get over what a godawful mess and pain in the ass it is. But on some occasions, for example, when the bleeding is not just inconvenient but metaphorical, when the blood flows from what feels like a wound—I am not pregnant, *and I am bleeding*—it is a shock.

All kinds of things would go through my head during this time. I am not getting pregnant because I used up all my luck landing someone as special as David for a husband. I am not getting pregnant because I made such a big deal about the whole thing before or because I am still not *ready*. I am not getting pregnant because I am too busy helping to take care of my grandmother; *she* is my baby. I am not getting pregnant because that is just the way of the world, some people do and some people don't.

We went to a doctor, an infertility doctor. At first I kept referring to him as the fertility doctor, until I realized it was *infertility*, as in, that's what he treats people for. But I preferred the other, with its vague voodoo appeal; if he had wanted to dance around me singing *sha sha sha* and sprinkling herbs, that would have been fine. Instead, he gave us a little pamphlet which he had designed to explain the course of our treatment. We would take all the easy, noninvasive diagnostic tests right away, in the hopes that our problem, that is, the one that wasn't caused by some kind of bad karma I had drummed up for myself, would have a noninvasive solution. *Invasive* being another relative concept, but I'll get to that later. The doctor gave me a routine pelvic exam, told us that he did not take insurance, and we were on our way.

Our first official test was the Post-Coital Test, where David's sperm was swabbed from my vagina the morning after and tested for signs of life, for pep, for mettle. The doctor showed them to me under a microscope, zipping around, frantic, directionless, wagging their tails. Like college dropouts, aimless but free. Expecting to make millions anyway. I thought them very beautiful and I was proud. *My children.* Next came the Quality of Ovulation Test, wherein my serial blood progesterone level was tested seven, nine, and eleven days after ovulation. Without sufficient progesterone the uterine lining cannot nourish the egg. If it adheres at all it will quickly slough off and die. That sounded very sad. People with this problem often miscarry over and over and do not even know it. They spot some time in the week or so after ovulation, and then they get their periods as usual and it doesn't occur to them that anything is wrong. Sitting in the doctor's office I wracked my brain. Had I spotted? I actually wanted to have spotted. What I wanted most was to have a definitive reason for not getting pregnant. The best you can hope for is that the doctor says, "Eureka! I found it! I'll just fiddle with this and tinker with that and you go home and take three of these and stand on your head reciting 'Ode to a Grecian Urn' and you'll be fine." There are many amazing remedies for infertility these days that didn't exist even five years ago. But you have to fit into one of the categories or you could end up like space junk, untethered, floating off into the galaxy, a wasted miracle.

But hey! I had a progesterone deficiency. I started using natural progesterone suppositories after each ovulation. That's as gross as it sounds. But I felt so proactive. And we didn't stop there, because infertility can be the result of a confluence of pathologies. So next we had the Sperm Anti-body Test. Evidently it is possible for the sperm to reject the vaginal mucus or vice versa. (Don't freak out here; if you want to have a baby you have to say words like *vaginal mucus* out loud or at least mutter them under your breath.) Hooray! We win again! David's sperm was creating antibodies. By the time they had swum upstream they had formed little protec-tive helmets that might be keeping them from penetrating the egg. So next came the Sperm Penetration Assay: Could the sperm penetrate the egg, under optimum conditions? That is, without their helmets? Why, yes. Yes, they could. Okay. We responded to this news with monthly IUIs. Those are intrauterine inseminations—sessions with the doctor and a high-tech turkey baster which he used to place the sperm directly at my uterus, avoiding my pestilent fluids. David would masturbate in the morning and drop off the sperm at the doctor's office on his way to work. I would show up about an hour later and after confirming that the sperm the doctor was offering was, indeed, David's, I would accept it from the doctor, like a communion cracker, sort of.

But we kept having sex, too. Because, as the doctor said, who knows?!

The fifth test on our doctor's prospectus was the Hystero-salpingogram or, if you were feeling zippy, the Hysterogram, for short. I was warned that some found this procedure painful and it would be a good idea to take Advil or Motrin before and after it was performed. What I was not warned was that this procedure was excruciating. A doctor, not my doctor but one he'd recommended, flooded my uterus with a dye solution and then x-rayed my entire reproductive tract looking for blockages. He suggested it might feel like slightly more intense menstrual cramps. Nice try. I knew what intense menstrual cramps felt like and this wasn't it. This felt like someone was ripping my insides out with a pair of needle-nose pliers and on top of it, the unavoidable sexual aspect of it, not sexual as in sexy but sexual as in involving my sex organs, was profoundly disturbing. It may be the most I have ever felt invaded. I sobbed hysterically through the entire thing. David could hear me in the waiting room and asked if he could come in and they told him no. Most people probably don't know this and don't need to know this, but the Greek root of *hysteric* means "suffering in the womb."

There's a point when you are starting to think about having children that you begin to notice other people's. And you say to each other, I want one of those. And you get excited. And then, oddly, the longer you try the more children you notice. In fact, everyone on the street is pushing a stroller and the ratio of pregnant women to non-pregnant women in your neighborhood is at least three to one. Everyone has a baby but you. Everyone *can* have a baby but you. All the other things your life is about aren't what it is about anymore. They all go away. What were all the things I used to do? I was in a lot of plays. I was in some independent movies, but then, everyone in the world has been in some independent movies. I still hadn't gotten on *Law & Order*. I wrote some stuff. A script, some essays, and some stories that I performed around. I went places with my husband. We went to movies and met up with friends and played tennis and went on ski vacations together. He's a terrific guy, of course, but he's no baby.

Everyone says that the best way to get pregnant is just to get on with your life and not worry about it, and that's what I tried to do. On the outside. But on the inside I did nothing and went nowhere. On the inside I was sick at the thought that I might not be able to have a child. I finally had a husband and security and work that I cared about, all the things I'd always considered the prerequisites of motherhood. And I was less woondangled and hublittered than I'd ever been. It never occurred to me in a million years that anything I was doing in my life was about becoming a mother but in retrospect, all roads led here. They would, I presumed, lead away again at some point. Or who knows, maybe not. But I kept coming back to the idea of Luck. I just wasn't lucky. Other people were lucky. Why were other people lucky? And to me, being unlucky wasn't just a manifestation of unseen forces in the universe; it was a personal flaw. Luck was something I somehow *failed* to have. Which by definition is contradictory, but it's so like me to seek blame and yet refuse it at the same time.

Every month I would take a blood test ten days after my presumed ovulation. A blood test could determine pregnancy several days earlier and with greater accuracy than a urine test, and the doctor wanted as much time as possible to prevent an early miscarriage. So one morning each month I would give blood at a lab on Eighty-sixth and Park and then the doctor's assistant would call in the afternoon with the results. In order to prevent any miscommunication, her greeting was this: "Hi, it's Suzanne, I'm sorry, sweetheart." That was it. Every month. And then, after a few bloody, disconsolate days, she'd call with the next month's instructions and we'd start again.

I had a choice. I could take a Trial of Fertility Drugs or I could have the Laser Laparoscopy and Hysteroscopy. The latter were surgical procedures to find and repair endometriosis and/or, if possible, address any blockage not revealed by the dread Hysterogram. I had neither the symptoms nor a history of endometriosis, so my doctor recommended the drugs. I panicked. Even the lowest dose of Clomid, my doctor's drug of choice, carried a 6 to 8 percent chance of twins. What a fertility drug does if you are not ovulating is to make you ovulate. If you are already ovulating, and I was, it induces the release of an extra egg or two, thereby upping the chances for fertilization. Yikes. Still, praying to the Forces of Good, or whatever, for one baby but not for two definitely did not seem right. I agreed to take the drugs.

I was freaked out by the Clomid. It made me feel very on edge and I was already on edge. I was on the edge of the edge. I spoke in terse, one- or two-word sentences. My head ached. Keep your eyes on the prize, I told myself. Or prizes.

Four eggs came down the chute. One was definitely mature, one might mature in time for ovulation, and two were teeny-weeny specks, useless, wasted things. That's the way with the drugs. To help the front runners I was given a shot of HCG, or Human Chorionic Gonadotropin, in the ass. HCG, a naturally produced hormone, triggers ovulation within a certain window of time, so everyone, the eggs, the sperm, and I, can all get to the theater for curtain, so to speak. We also did two IUIs about twenty-four hours apart; David came home at lunch and we looked at a copy of *Jugs* together. Afterward, I started again with the progesterone suppositories.

Nada.

When my period finished I took another round of Clomid,

this time with no misgivings. Bring on the babies. The more, the merrier. It had been a year since we started trying to conceive. That probably doesn't seem like long to some people, but since I have a self-destructive habit of calculating my life in dog years, it was an eternity to me. We had all but worked our way through the doctor's little pamphlet. He didn't even list In Vitro Fertilization as an option because everyone knows it's the last one. The last-ditch effort. I did not want to end up there, at the end of the line, with nothing else to try. At the same time, the doctor had yet to prove that conception was impossible using plain old sex. David and I could still keep trying after all the extraordinary measures had been exhausted. We would be like the inhabitants of *Gilligan's Island*. We would stand on the beach and shout at ships on the horizon until Kingdom Come or we were canceled.

Our doctor was a very fine doctor. We'd heard stories of couples he had helped who had been told by other infertility specialists to adopt. Perhaps he was like the doctor in Orange County who was caught impregnating his patients with his own super sperm. I tried not to care. The man was reasonable looking, and certainly he was very smart. He was a doctor. The only thing I didn't like about him was that he talked too much. He chatted through every procedure, including the inseminations, most likely with the intention of keeping me relaxed. It had the opposite effect. He and his nurse would quiz me about current Broadway shows while he doused my uterus with David's sperm. I don't like it when

David says anything other than "Oh, Cindy"; why would I want to talk theater? But I couldn't find a graceful way to ask them to respect the sanctity of the moment. It seemed ungrateful. So I brought in my Walkman. I put my headphones on and closed my eyes and I guess they got the message because all I could hear was the Boss.

I went into myself, into my body, *Fantastic Voyage*–like. I swam north with the sperm, single-minded. We were heading upstream to spawn. The doctor had actually praised this particular batch, so we were pretty pumped. We would save the species from extinction or die trying. We courted the hell out of those eggs. We took them out to dinner and held hands with them under the table. We walked twenty-five blocks in the snow and talked as if we'd known each other forever. We bought them flowers at the all-night deli and then slow-danced to "Jungleland." We ate Häagen-Dazs from the carton watching Dana Andrews obsess over Gene Tierney in *Laura* on the Late Late Show.

Then we made our move.

Once again I went for my blood test. And once again I spent the afternoon in stone cold dread of the inevitable phone call from Suzanne. Things had gotten to the point where I could not imagine any answer other than no. I'd stopped envisioning my body as being capable of anything at all, much less conception. Nothing ever happens the way you think it will. When David proposed to me, I had to ask him to repeat himself because I hadn't been listening. I had been thinking about my grandmother, Lil, who had just died, and my father, who had just had a heart attack. The whole thing, the romantic restaurant, the roses, the funny, crappy fourteen-dollar ring from Weber's Closeout, all smooshed by in a blur. Most people I know are capable of giving highly detailed, minute-by-minute accounts of the defining moments of their lives, and I'm always like, "What did you say?" That is how I felt about trying to become pregnant. Like the moment of truth had somehow come and gone without my quite knowing what had happened. It was supposed to be so easy. Like on TV. "Remember when we made SusieJustinBrittanyMax?" "He touched my deepest fla fla." "I don't know how but I just knew." I'd laid there with a pillow under my butt and my legs in the air and I knew nothing. I'd taken my temperature and put my fingers inside myself to gauge the quality of my vaginal mucus (stretchy like egg whites = fertile; tacky like old jam = infertile). I'd performed countless at-home ovulation tests and I'd gulped Robitussin, because I'd heard of women getting pregnant during bouts of cold and flu and that they credited the mucus-thinning agent in cough medicine. I'd gone to the best doctor I could find and done everything he told me to do. I'd wished on eyelashes, on chicken bones, on candles, on stars. And still, I knew nothing. And I'd come to the conclusion that those home pregnancy kits were a joke, a marketing scheme. They would never turn the other color or show two lines or three dots or a smiley face.

So I waited and finally Suzanne called. She started talking about how I'd brought my Walkman in and something else, but I didn't hear her because I was listening for the word "Sorry."

"Well, it must have been the Walkman," she said.

Where was "Sorry"? Come on. Sorry, sorry, sorry.

"What?" I said.

"It must have been the Walkman."

"What?"

"It must have done the trick because you're pregnant."

I just kept saying, "What, what? What did you say?" like I always do. And she repeated the whole thing again and I cried loudly and said, "Oh, my God," thirty times.

"Call your husband," she said.

I did. I called David and cried, "We did it!" That was the best I could do. I knew I was supposed to show up at his office with a bottle of champagne or FedEx him knit booties or casually refer to him as Daddy-o over a candlelit dinner. But I had no self-control. I was utterly incredulous, as if a dinner plate had spoken to me in French.

Dumb luck. That's all it was. Or, who knows, Grace. Divine leniency. This morning when my son saw me his face did this: Is that you? It *is*! It *is* you! Oh hooray! Hooray! HOORAY! And now, here we are riding up and down the elevator in Bloomingdale's looking for pink shoes.

mountain men

———◆———

IMAGINE, if you will, that it is late at night, maybe you've just come home from some jolly event, some dinner dance or cocktail party, and you're feeling happy and relaxed. You're getting ready for bed, grateful to be taking off your bra, when you hear a slight rustling behind you, so slight you might have made a mistake. Maybe it was just the wind in the curtains.

If you had curtains.

Suddenly, as if unleashed from a cage, It spirals frantically towards you. You scream, "Oh, God," and run from the room, hoping against hope It does not follow you. Your husband springs into action. He knows what to do; he has been through this before. He must go into the bedroom, shut the door, and kill the moth.

This is why I will be a bad mother. I am pathologically afraid of moths. I can't stand their insane flapping, their arbitrary, freaked-out flight patterns. Their fuzzy, dingy bodies and papery wings disgust me. They are like wads of used tissue on acid. How they get into our apartment on the thirty-third floor is a mystery that haunts me. But they do. And sometimes they prove so elusive that I will walk and sleep in terror for days after their initial sightings, further tortured by the knowledge that one day, maybe years from now, I will find them petrified on surfaces upon which, given their utter incapacity for self-control, landing would seem inconceivable. Book bindings, the edges of drawers, the sides of poster frames. In the folds of shower curtains, on the cuffs of sweaters. I will jump at their dead selves, still horrified by their powdery, innocuous existences, and I will wait for David to arrive home and remove them.

I should admit that I have other mostly morbid, unfounded fears, like of avalanches and Legionnaire's disease, as well as perfectly normal fears that are remarkable only in that I dwell upon them for inordinate amounts of time and with unhealthy zeal until they have stopped being normal and become morbid and unfounded. I have tried to trace the history of my dread, starting with fear of adults and culminating with fear of being shot in the head by a sniper. In between, I have documented fear of being knocked over at recess by a big girl named Heather, fear of getting into a fatal argument over the TV remote, fear of Drano.

And now, of all things, I am a person with a child. The idea of having your own is such an abstraction until the day arrives when you do. And then it is terrifying. An entire new assortment of fears descends upon you and you realize they will never ascend but rather multiply exponentially until you are carted away to the place where crazy people ride giant tricycles that your mother has requested *she* be sent to when and if the time comes that you do something so stupid and dangerous that you finally drive her out of her mind for good. Oh, my God, where was I? Oh, yes. In the first weeks of parenthood you are so completely wasted that you just feed the baby and change it and hold it and do all these things almost without thinking about it. And then it dawns on you: It's alive. The poopy thing is alive and it is up to you to make sure it stays that way. Whatever you do, don't drop it on its head. Don't spill hot liquids on it. Don't drink hot liquids. Be careful of its floppy neck. Be careful of its nose and mouth. Keep those airways free. Don't forget that it exists when you go to the bathroom. Don't wake up in the middle of the night and think for a second that it isn't there, next to you, in the bassinet. Unless, of course, you fall asleep together after nursing and it is actually in your bed. Don't roll over on it.

Once upon a time my husband and I called each other from work to rehash the sex we had the night before. Now he calls me or I call him and this is what we say, furtively: "I

have a bad story. I'm just telling you because it is a warning."
Ugh. It is the worst of all possible kinds of gossip; it is the
dissemination of the Tragedy. Someone else's tragedy, that
is, often someone we don't even know or have met only
briefly or is a friend of a friend of a friend.

And then I say to David: "No Fritos until the baby is four,
no, five. No nuts and no hot dogs and no doughy bread and
no peanut butter and no popcorn and no grapes. And no gum
and no small hard candies. No Pez."

"No roughhousing," I say, that night, when the baby is in
his bed. "You always have to be in control. The baby can get
out of control but you can't. No jumping, no flinging, no
tossing. Don't trip."

I wake David at 3:00 A.M. to say, "He gets his own seat on
a plane."

I call my friend who also has a baby. I tell her our most
recent stories and then throw in some more accrued over the
months. Stories that I have deposited in my bank of terrible
stories, the account to be drawn upon and its filthy currency
spent whenever I am feeling nervous or helpless, whenever I
need to make a strong point in favor of doing things my way.
The careful way. The boy-in-the-bubble way.

Recently, my niece broke her wrist while she was on vacation in France with my brother, his wife, and their other
daughter. She fell off a swing. Let me say first that I was not
in France. And yet I have replayed the scene, or my render-

ing of it, over and over in my imagination. I take nothing in stride. How could it have happened? Wasn't anyone watching? She herself told me about some Charging Wild Boars. What about them?

This is how it goes, late into the night. Clearly, there is an ineluctable crack in my psyche which will eventually require surgery, or drugs, or both.

A couple of years ago, as David and I rambled around our favorite store, a hiking-and-camping-supply store in Vermont, considering camping gear—tents, stoves, water-purification systems—I happened to open a book with a title that was something like *Bear Attacks*. What possessed me to do this is a question I have been asking myself on every hike and campout since. Did I need to read about people having their arms ripped off by ferocious bears? Did I need to read about people whose heads were mauled and torsos chomped and bones stripped bare? Did I need to hear them tell their tales themselves, in first person singular, an indication not only that they lived at all, maimed, scarred, in all their armless glory, but that they were semiconscious during the attack? If David thinks that we will take our baby camping with us next summer, he can do it with his new wife.

Are there really bears out there? Yes. Would David and I take the baby hiking in serious bear country? No, of course not. And isn't a rousing chorus of "I'm 'Enry the Yeighth, I Yam" in a bad cockney accent sufficient to scare the occasional bear off? Besides, all bears really want to eat is your Doritos and your toothpaste. So, honestly, is there anything to be afraid of?

Yes. Mountain Men.

Mountain Men are men who live in the mountains and occasionally come down and kidnap female hikers. At least this is my understanding of them. They have long, matted beards and wear ancient Wrangler jeans and cheap flannel shirts and furry vests. They look like a countrified ZZ Top. They go too long without the company of womenfolk and litt'luns, which is what makes them so dangerous. Or so I gather.

Sounds pretty stupid, doesn't it? I think so.

My mother, like any really good mother, spent a reasonable portion of her life protecting her children from danger or at least trying to. She made me put one of those orange flags on the back of my bike and yelled at me when she passed in her car after I'd ditched it in the woods. She made me wear wool hats to high school despite the certainty that I would spend an agonized day with static hat head. She made me wear a Medic-Alert bracelet so, God forbid, any paramedic peeling me off the pavement would know I was allergic to penicillin. And her vigilance has not wavered with my adulthood. She still phones with hurricane alerts and migraine medicine updates and her newest clarion call: computer virus warnings. She hates that I take the subway, that I run in the park, that I visit friends in what she would call iffy neighborhoods. Ironically, I have pretty successfully repressed many of the everyday bugaboos associated with city life. I ride the subway at all hours. I walk dark streets with aplomb. I act unafraid and I pooh-pooh my mother to the point of insult: "Take a cab after midnight, are you mad?" Because what I really fear is the escaped mental patient, the rogue wave, the freak accident (*The chandelier just fell, just like that!*). I am more apt to think someone is following me on a nature trail in Montana than down an alley in Hell's Kitchen.

So is all this her fault? Probably not. In fact, in comparison to me I think she was a little, I don't know, criminally negligent. Even now, her blasé attitude toward Drano irks me. And in some ways she is my exact opposite. Her fears are for the most part founded; I *should* have used the orange flag. Mine, on the other hand, are baseless; they are the spectral imaginings of a lunatic mind.

Maybe this is a thing about control. Maybe I just want to be on my toes, to be prepared for anything. And perhaps if I were to be hypnotized, which I won't be, but if I were, it would come to light that my illogical, some might say, ridiculous fears are actually the conscious manifestations of their more logical but repressed counterparts, i.e. that it is just *easier* to fear a Mountain Man, since I am only at risk from one a few times a year, tops.

One morning, there is a penny in the baby's poop. Who put it there? Is it supposed to be a joke? Did the baby somehow pick up a penny and tuck it into his diaper? He has such a crazy sense of humor. You wouldn't believe the things he laughs at. Anything on my head is an automatic gut-buster, as is the sound I make when I cough. I could be dying of consumption and he would be howling with glee. But, really, what's the story with the poopy penny? I cry several times throughout the day, thinking to the edge of the unthinkable, of how it might have been different. My husband and I retrace our steps. We can't figure out when it could have happened. Which is worse? To know exactly when you weren't watching closely enough or not to know?

I call a couple of friends, friends who have children. They gasp. I say to each of them: "No, you're supposed to say, 'A penny, why, that's nothing! Why, I once found a matchbox car/tennis ball/commemorative silver dollar! Why, a diaper is nothing more than a salvage yard!'" But everyone is stunned and I am sickened. "When did it happen?" one woman asks. "I don't know," I say. "Some time over the last couple of days," I say. "No, no," she says. "It takes a week or two for coins to work their way out." "Ah," I say. "Then it was that awful moment a week and a half ago at our friend's house in Vermont when we thought the baby had maybe picked up a sharp little corner of a tortilla chip. He was cry-ing and coughing so we let him work it out himself—that's what they tell you to do—and then he was okay. It must have been the penny." "Yes," says the woman, a mother, too. "Yes, it must have been."

I get an e-mail from a friend. *It was a lucky penny,* she writes.

Yes. Oy.

How do I go forward without a lobotomy, you ask. That's the first question. I don't know the answer to that but I'll ask my mother, she might. Next question, and more to the point: How do I raise a child not in *this* world, but in *my* world? There's no way to answer that except to say that it is too late to turn back now that I own a Tiny Love Gymini Deluxe 3-D Activity Gym, an Evenflo Excersaucer Junior, and a Graco Doorway Bumper Jumper. So when we are home alone and the moths are doing their demented dance, I will muster up some courage. I will. Or we'll wait together in the bathroom until help arrives.

megrim

—◆—

I started having headaches in fifth grade. It may well have been a result of the experimental classroom situation I was in. Twenty ten-year-olds spent the entire day in an enormous room with just each other and two teachers, one of whom looked like Susan Saint James. Or it might have been the New Math. Or perhaps it was our progressive curriculum's constant demand for self-determination in the form of Special Projects. Around the time the headaches began I was working on an idea for a project to write a murder mystery with farm animals as the main characters, and the plan was to actually make life-size stuffed animals who would stand around me while I read the story to the class. A person's mother would have to use a lot of brown felt to make even a smallish horse. When the big day arrived, not one of the animals, not even the duck, could stand on its spindly, polyfill legs. The headaches continued.

My mother took me to Norwalk Hospital to have an electroencephalogram, commonly known as an EEG. They—when you are ten years old all medical personnel may be justifiably referred to as They—stick fifty or sixty pointy electrodes just beneath the surface of your scalp and measure your brain activity. Your brain waves. The ebb and flow. High tide, low tide. If there is a disruption, something that upsets the regular poom-poom of the surf upon the sand, a sloop wrecked upon a coral reef or a storm a-brewing, or, *say*, the presence of a large unidentified mass, the EEG will detect it. Of course, an EEG is only prescribed when indicated by, *say*, the sudden onset of blurry vision followed by the horrifically painful sensation that your head is about to cave in.

There I lay, whimpering, sprouting wires, Medusa-like, while they recorded what I presumed were my ten-year-old thoughts. Would they learn that I was mortally afraid of Heidi Holbrook, who was known to pin fellow classmates to the blacktop and pull up their dresses, looking for what? Or that I often threw my American cheese sandwiches in the trash, once provoking a teacher to give the entire cafeteria a lecture on starving children in China? I had no idea what could be wrong with me; I didn't read *Death Be Not Proud* until seventh or eighth grade.

* * *

Here's what a migraine headache feels like. First, it feels like you think you have a brain tumor. It doesn't matter how long you've been getting them, that morbid notion is always loitering there in the back of your mind. So that's first, the Ridiculous Fear. Then there's the Justifiable Dread, humming harassingly in your ear as you prepare for every major event of your life. Will you get a migraine on the class trip to Mystic Seaport? In the middle of performing the spring musical, *Damn Yankees?* Will you get one while you're taking your SATs, during your twenty-first-birthday party, at your college graduation? Will you have to lay down on the floor in the hallway outside your wedding reception? Will you forget to take your migraine medicine with you and have to canvass friends and strangers until you have accumulated at least eight Tylenol or Advil? Then comes the Moment of Truth. Do you really have the blurries? Or is it just a head rush, or snow blindness or photosensitivity or a floaty in your eye?

About 15 percent of migraine sufferers experience what is referred to in the medical world as an "aura" before they feel any pain. This is not the Dionne Warwick kind of aura. My aura, what I call the blurries, is actually, in migraine-speak, a "visual disturbance" during which mini-bolts of lightning appear, slowly at first, and then with increasing intensity, until everything in view looks like a TV after the national anthem has been played. It's pretty disturbing. In the first moments of their appearance the blurries are accompanied by an intestinal groan, a sort of lamenting in the bowels, which may itself be either a feature of the headache or the physiological manifestation of Utter Dismay. Of course, things could be worse. Other migraine aura options are temporary paralysis of one side of the body and slurred speech.

When the blurries start, it's as though everything is closing down and closing in; the migraine is the dark shadow in a Scooby-Doo cartoon and you are the cringing Gang. The blurries usually last about twenty minutes, then, as the flashing recedes to the periphery of my vision, the pain begins. It is almost immediately intolerable. It defies description, but let's give it a shot, anyway. It is at once sharp and plodding, exact and diffuse. It is so large that it is difficult to place exactly where it is worst. It throbs behind the eyes and at the temples; it grinds at the base of the skull. Sometimes you are doubled over with terrible, bubbling nausea. It is as if a bear has chomped on your head and is trying to dislodge it from the rest of your body, and you are simultaneously sickened by the thought of your decapitated self.

In those first years my mother would pick me up from school and put me to bed with the lights out and the blinds pulled down. I would lie in the dark, keening, praying for sleep. Later, when my brother and father came home, there would be a lot of shushing and creaky-staired tiptoeing. By high school, if my mother couldn't be reached, my friend Mary Beth was permitted to leave class and drive me home in the car I shared with my brother. She would spend the rest of the day watching TV at my house. By tenth grade I was taking Cafergot, a mixture of ergotamine and caffeine intended to constrict what were then believed to be excessively dilated blood vessels. The trick with Cafergot was to take it before the headache kicked in and then to get to sleep before the caffeine kicked in. The main trouble with Cafergot was that first, it was not a painkiller, and second, it was not a painkiller. My next prescription, Ergostat, consisted of a tiny pink pill, to be placed under the tongue where it would slowly, chalkily melt. It had a vague peppermint flavor and was notorious for inducing extreme nausea. To this day certain mints still inspire horror, and the words *under* and *tongue* uttered in close proximity have the same Pavlovian effect on me that *nest* and *egg* had on Albert Brooks in the movie *Lost in America.*

Another unfortunate side effect of migraines is that they make you kind of a bummer to be with. If you are out with friends at night or on a trip or somewhere very inconvenient like on a ski slope, everyone or at least someone has to stop what they are doing and wait for you to recover enough to continue. Everyone has to be solicitous and concerned and not act annoyed that the works have come to a screeching halt. So you get to trying to hide your headaches. You say, "You guys go ahead, I'm going to skip this run and take a breather/admire the view/examine my bindings for a while." And you hope that someone sidles up to you and says, "Yokay?" so you can whisper, "Migraine," and then at least one person knows you're in potentially desperate straits.

No place is sacrosanct. I often get them in the middle of the night and the Me that I am in my dreams will get the blurries, too, interrupting the dream-event at hand to go looking for painkillers, usually aspirin, not anything stronger or more current, in that same way that every time I dream of home I dream of the house I grew up in. Then the headache will begin and I will wake up and fumble in a muddle of déjà vu for my Tylenol 3. I will stumble to the kitchen for an ice pack and a dish towel and I will affix the former to my forehead with the latter. It is possible I will groan as I await relief. When my husband wakes in the morning he will find the remnants of my dead-of-night soiree: a defrosted rectangle of blue liquid in a plastic puddle beside my head and the towel over my face—defense against the dawn. Sometimes I will wake David and let him/make him take care of me, a luxury after so many years alone. Maybe that is one of the best things about being married.

There is an idea, I think, propagated by *Masterpiece Theater*, that people who have migraines are whiners or shirkers or both, or else they are dangerously, unbearably fragile. Women on *Masterpiece Theater* are always claiming "The Megrim" in order to be excused from some unappealing activity, often one involving sex with their husbands. Conversely, a megrim was also an excellent reason to stay home from, say, the hunt, and tryst with one's lover. Unfortunately, some seemingly authentic sufferers, it would turn out, were actually the victims of undiagnosed brain tumors. Anyway, back to the point, or at least, the point I am making here, as opposed to the larger point, which *is* about brain tumors: migraines rarely occur in the midst of even extraordinary pressure. I used to have a boss who, at the height of the day's tensions, would exclaim, "This is giving me a migraine!" The poseur; it drove me crazy. Migraines are *post*-tension headaches. They happen when the buildup is over, when you finally relax. They like to strike when you least expect them. While this reasoning seems paradoxical, it is not. Most of the worry, most of the tension associated with an important event occurs as the groundwork for that event is laid. What's a good example? I know one: childbirth. You would think the months and weeks and days leading up to the big event would be chockablock with headache-inducing stress. Picking the pediatrician, buying some diapers for the thing you hope doesn't turn out to have two heads and one foot—that can be very worrisome. Yet, not a megrim in sight. But activity is the antidote to anxiety. Once the ball starts rolling there is a necessary letting go, a release. My water broke at ten o'clock in the evening. We went to the hospital, and no sooner had they given me Demerol to help me sleep through the night than I got a migraine. On Demerol.

Of course this is just one theory. But it is, among others, roundly accepted in migraine circles. Other theories have to do with food triggers, such as red wine, chocolate, aged cheese; hormonal changes, particularly in conjunction with menstrual cycles; and my favorite: weather. I am a human barometer. Sometimes my migraines begin precisely as a storm is moving in or out, as a dry front blows away a wet front or vice versa, or a heat wave is broken by a high-pressure blast from the Gulf Stream, whatever the hell that really is. Mention El Niño and most migraineurs will blanch and shudder.

I recently ran into a woman I know on the street and she told me that six weeks earlier she'd had a benign brain tumor removed. In February she'd woken up deaf in her left ear and after a lot of testing by an Ear Man she was sent for an MRI. A tumor in the center of her skull was depressing an aural nerve. But they got it out and now she's fine. The kind of tumor she had was a one-in-a-million thing. "Well," I said, "after something like that you're pretty much in the clear." "You'd think so," she said, "but actually when I was five I had brain cancer, and that was a one-in-a-million thing, too, and the two are totally unrelated, so go figure."

I'm figuring. I'm figuring that maybe she had both *her* one-in-a-million thing and *my* one-in-a-million thing and now *I'm* in the clear. On the other hand, it makes you think that you're never in the clear. And the truth is, bad things befall the same people over and over again all the time. The same way that disasters often happen in clusters. But, still, brain cancer and a brain tumor? What's the deal with that?

I thought about this woman when I woke up one morning and my left eye was tearing. Why was my left eye tearing? Because I had a brain tumor and it was pressing down on my tear duct, that's why. What else could it be? All right, maybe I had a little cold in my eye. Or maybe part of me was just trying to enjoy the good life while the other part of me was heartsick over apartheid. Or maybe the right side of my brain was sad that I've never been good at math. My eye cried for over a week. I felt like the clown with the single tear. Although, why is he dressed like that if he's feeling so bad? Anyway, I was concerned about the condition's primordial source, if you know what I mean, so I went to the eye doctor. She checked my lens prescription and sent me on my way. Undeterred, I made an appointment with a playwright I know who also happens to be a neurologist.

Actually, I was going to see him anyway. Besides the fact that I knew Jeff would be particularly sensitive to the enormous dramatic potential of a cold in the eye—it's good to have a friend who is both a healer *and* an artist—I'd recently had a very bad run of migraines. Whenever there is an escalation in the frequency of my headaches I go on the lookout for a good opportunity to check on the status of my brain tumor. It is probably still too small to show up on a CAT scan or an MRI, so it was my plan to have Jeff perform what I refer to as the Lindsay Wagner Inert Arm Test. There was a TV movie many, many years ago starring Lindsay Wagner as a woman with a brain tumor. She had been getting headaches and having dizzy spells and when she went to the doctor he asked her to close her eyes and hold her arms out straight in front of her, perpendicular to her body. Lindsay closed her eyes. One arm rose and one did not. The doctor asked her if she was still holding both her arms out and she answered confidently that she was. The next time we saw Lindsay her head was encased in a helmet of white surgical gauze and her memory was kaput. Could her husband still love her? Could she love him? I'm not sure but I think I remember that the new Lindsay and the doctor fell in love. Hmm.

Or maybe it was Elizabeth Montgomery.

Jeff performed a variety of other coordination tests. With my eyes closed I raised my arms (both of them) and I touched my index fingers to each other and to my nose. I followed Jeff's finger with my eyes to various points around my head. Then he felt my skull for telltale lumps, and assured me that the one I thought was a brain tumor was actually the tip of my cranium. He also examined my thyroid and popped my knees and elbows with a small hammer. I was ever so slightly self-conscious, being that he was also a friend and colleague whom I have known since before my marriage, and I blushed embarrassingly when he patted the examining table because I assumed he meant for me to lie down, which he didn't. I don't know why this should have been embarrassing, but it was. Maybe I was overzealous for an examination of some sort.

I recounted my history with migraine medicines and, as I

expected he would, Jeff asked me why I hadn't tried the new sumatriptan drugs like Imitrix or Zomig. I told him that I hadn't seen the point when I knew I wouldn't be able to use them while I was pregnant or nursing, and I could still take my trusty codeine. Besides, who doesn't like a narcotic? Codeine is a delightful drug as long as you don't take it on an empty stomach. How hard is it to have a little nosh while waiting for the blurries to subside? But Jeff told me that the new drugs weren't painkillers, they actually interrupted the relay of the message that your brain is sending to your head to stab itself repeatedly for the next five hours. Would they, I wondered, interrupt any other messages? Like the one that makes me furious when people don't use coasters on wood furniture? Jeff gave me some sample pills in a brown paper bag. The next week David and I went on vacation, and I got three migraines in seven days, which proves, once again, my theory that migraines eschew the daily grind. Unfortunately, I'd forgotten to bring the new drugs, so I spent a good part of the week doped up on codeine. Not the vacation I was looking for, but not bad, either.

I feel I am at a crossroads. How much worse can this get? Or rather, more importantly, what did I do to deserve it? My personality is already so prohibitive. How many obstacles to a relaxing time am I supposed to have to surmount? An article I read recently suggested that it is unclear that anyone has ever been cured of migraines. I suppose this means that I am

going to have to cure myself of everything else. All the pho-
bias and neuroses and unhelpful quirks. But the two go hand
in hand, don't they? Headaches have always been harbingers
of ill. All I can do at this point is hope to God I never have
cause to utter the words, "We thought they were migraines."

Years ago, I brought my college friend Mark home to
Connecticut for the weekend, and at dinner on Saturday my
parents and I regaled him with Kaplan lore, or what little
there was of it that would be of interest to an outsider. I told
the infamous Dog Story, which begins on a melancholy note
with the death of our first dog, an incorrigible cocker spaniel
named Jasper, who fell through the ice on the river behind
our house and drowned. It meanders over six years of tearful
pleading on my part for a second dog, to no avail. Then
there's the stunning twist when, the summer after I turned
eleven, my parents announced that it was their intention to
get another dog. We got one. End of story. Light applause.

What changed their minds? asked Mark.

Yeah, what changed their minds?

"We thought you were sick," my mother said.

"What do you mean sick?" I asked.

"You were having terrible headaches and had to lie down
in a dark room. James Stewart's wife started getting terrible
headaches and had to lie down in a dark room and then
dropped dead two weeks later. (For years I thought she
meant the actor James Stewart but she didn't. There's some

other prominent James Stewart. Someone only my mother would know of.) We didn't know what to think. We were worried."

"So you got the sick girl a puppy," I said.

"Yes."

"Did you think the puppy would make her better?"

"Well, we thought it might help somehow."

"Or did you think, get her a puppy before it's too late?"

"That, too."

passing

DURING all of seventh grade I went to two bar mitzvahs. There were only what seemed like a handful of Jewish kids in my school, and there was no synagogue in our town. A few of us went to Sunday school at the Reform temple in the next town over. To each of these bar mitzvahs I wore the same outfit I wore on airplanes: a navy blazer, a kilt, a button-down shirt, knee socks, and loafers. I was seated with the bar mitzvah boy's Christian friends.

When my brother and I were young we got Christmas presents in addition to Hanukkah presents, so we wouldn't feel left out. Unfortunately, the result of this seeming abundance was that we got a lot of little, meaningless gifts, as opposed to one very large and excellent gift. My all time worst Hanukkah present was a giant yellow pushpin that was supposed to hold down all the stray little pieces of notepaper with people's numbers and such on it. I was maybe twelve or thirteen at the time and I am pretty sure I asked my mother, "What the fuck is this?" I didn't get in trouble, because it was a legitimate question. My brother and I watched the Christmas shows—*Frosty, Rudolph, The Grinch*—all of which usually aired during the eight nights of Hanukkah.

There are no Jews in Whoville.

Every Christmas Eve I was invited to one or another friend's house for the tree trimming, and occasionally my family accompanied friends to midnight mass. For the music, my mother said. She meant that, and I understood. I love Christmas carols, I love their soaring melodies and peace-promoting messages. I love that they are in English. *"Hark, the herald angels sing!"* Is there any more glorious beginning to any song, aside from Neil Young's "Heart of Gold"? So it's about Christ, so what? I'm not even totally sure what "Heart of Gold" is about. And how about those standards? "Have Yourself a Merry Little Christmas" and "White Christmas." George Gershwin never wrote "I Love You *More*, on Yom Kip*pur*." Why should he have? Who wants to write a song about a bunch of people sitting around hungry all day? And although every Jew I know claims that Passover is a favorite holiday, nobody seems particularly inclined to gather around the piano and sing the song about the two zuzim.

Still, there was never any doubt in my family circle that we were Jewish. We acknowledged our faith at high holiday services and seders and family meetings, which were something my father's large family insisted on having every year, as though there were some kind of annual political agenda to be addressed. The longest discussion usually revolved around where next year's meeting was to take place.

"It should be at Phyllis and Milton's!"

"What? I still haven't seen Jack's house in Connecticut!"

"Oy, Connecticut's so far."

I met my Jews at the Jewish country club my parents belonged to and at my Jewish summer camp. Our family friends were mostly Jews, left over from our life in Manhattan, and some of them had summer houses in Connecticut. My mother never actually wanted to move out of Manhattan in the first place and referred to our house as Misery Manor. She joined a bowling league and was a Cub Scout den mother and to this day I am not sure she has completely recovered from either. My brother was not bar mitzvahed at thirteen because it seems the bar mitzvah classes conflicted with his busy Ultimate Frisbee schedule. At sixteen, either a renewed interest in his Jewish heritage or a pressing need for cash inspired him to rethink this decision, and he and my father, who couldn't remember if he'd been bar mitzvahed or not, decided to study together. On a Saturday morning in the spring, we celebrated their accession to manhood with smoked salmon on pumpernickel rounds and all-white-meat chicken salad from William Poll. This was our version of Jewish food. Not a pickle in sight. No one was even remotely concerned about whether I was bat mitzvahed, so I wasn't, which was fine with me. First of all, I would have had to give up field hockey, basketball, and the spring musical, *Finian's Rainbow*, and second of all, at the time, it was antithetical to the cause. The cause being my assimilation into the non-Jewish world.

Actually, that is not exactly right. I wasn't trying to assimilate. I just *did*. I have always passed. In my town there were so few Jews that unless you were dressed like the Hasidim, it just didn't occur to people that you might be Jewish, although now that I think of it, most of them had probably never seen the Hasidim. And if your name didn't start with *Silver* or *Gold* or end in *stein* or *berg*, you were pretty much in the clear. I had a fair complexion and blue eyes and light brown hair. It didn't occur to *me* that I was different, either. Or it didn't seem as if there was much of a difference, anyway. Only once in a while I would get an inkling that I wasn't all I was cracked up to be. One friend's mother, when I mentioned that I wouldn't be in school the next day because it was Yom Kippur, said, "Oh, you're *Jewish*," as though I had just revealed some terrible secret, like my father was a transsexual.

I never heard a girl called a JAP until I was seventeen. There weren't enough Jews in Connecticut for word to travel overland from Long Island. For a long time I honestly thought it had something to do with the Japanese.

The culture shock I experienced during my first months of college was a result not of being on my own at a big-city school, but of seeing so many Jews in one place. We were everywhere. I was both mesmerized and repelled. The girls were warm and quick to make friends but the boys all seemed to be overdressed. They wore khakis or new looking jeans and loafers and Italian-made color-block sweaters. I immediately gravitated towards the grungy, WASP-y jocks. Some of my friends would only date Jewish boys, even at eighteen, just on the outside chance that they would fall in love and want to get married. They didn't want to give themselves even a chance to fall for anyone else. It was a notion they grew up with, that their parents instilled in them. My parents pretty much left me to my own devices, possibly because I apparently didn't have any devices, but also because they wanted, above all, for me to be happy. I know that if I had come to them and said "I love a Ubangi," they would have welcomed him with open arms.

In college, I felt that people had to rethink me when they found out I was Jewish, that I was more exotic, more complex. I looked like a nice Christian girl but I got to be Jewish, with all that that implied: the liberalism, the wit, the intensity that is a result of our collective indignation, and, of course, the classic Jewish superiority complex. Fraternity boys sometimes told Jewish jokes in front of me and as soon as I heard where the joke was headed I would start furiously plotting my unveiling. How would I announce myself? Should I cut right in? Should I wait and see who laughs? Should I . . . and by then the joke would be over and if I got anything out it would be a weak sort of "Um, I'm Jewish, you know," a peep beneath the blare of Duran Duran. I saw myself as a warrior in a kind of socio-religious war, but I was too shy to let my righteous anger be felt in a meaningful way. An argument between two boys I knew ended when one hurled: "You're such a Jew!" across the crowded cafeteria. I stormed over to him and said, "Why do you think that is an insult? Would you be insulted if I yelled 'You're such a Christian' at someone?" He looked at me like I was speaking in tongues. I wanted to be Jewish, but I also wanted to be like everyone else. I didn't like carrying the burden of two thousand years of persecution around when everyone else looked so fancy free.

And maybe I had some idea that if people got to know me and found out that I was Jewish, then they would think to themselves, "Hey, if *she's* Jewish I guess I must have been wrong about the whole lot of them all these years." I'm like one of those okay Jews. Not too Jewy. Which can backfire on me. I took my son to a class at the Ninety-second Street Y, a very Jewish institution, and the other Jewish mothers didn't talk to me at all. I wanted to say something offhand, throw in a Yiddishism, just to let them know, but I know very little Yiddish. So I have learned to sneak the words "Yom Kippur" into a conversation just to make sure people are clear about things. "That's a beautiful watch. A Christmas present from your husband? Mine got me one just like it for Yom Kippur." "I'd love to see pictures from your Caribbean vacation. I have some pictures from Yom Kippur. Do you want to see them?" And I am finally ready to do battle, not just in the name of Judaism, but in the name of anything. If therapy did one thing for me it pushed my anger from its longtime residence in the space between my ears and sent it rocketing out my mouth. At one time, if a car cut me off as I was crossing the street, the event would provoke a furious interior monologue about where the driver had to be that was so *important* that it was worth running me down. I would fantasize about confronting him, perhaps from my hospital bed or in a dramatic courtroom showdown. Now, I will just kick the back of the car as it makes the turn and yell "Asshole!" What a nice example I set for my son. I am so primed for an actual confrontation that David sometimes whispers, "Choose joy," if I seem to be building up a particularly frothy head of steam. Where this comes from, I don't know. Perhaps it is all those years as a good girl. Or perhaps it is what happens when you finally wrest your Jewish self from Connecticut.

And yet, in that I have not been entirely successful. I married a Jew with light brown hair and enormous blue eyes and, like most of us, two slightly different profiles, one sort of

Jewish, one not. This affords him a certain flexibility, if only in my diseased mind. And evidently we have named our child wide of the mark. Jewish people we know keep telling us that Jews are not named John, but Jon, for Jonathan. These are Jewish people who name their children Porter or Layne or Brooke. Fuck them. I like John. It is a strong name, a solid name. I can't help it if all of the popes like it too. I have loved men named John. I have never loved a Jonathan, never. One tried to kiss me once and it was horrible, all wrong.

As it turns out, John is as blond and fair as a California surfer. Even our rabbi gave him a second look. (May I say that our rabbi is a hunkalunka and makes being Jewish a double blessing?) It is clear to me that part of my responsibility to John will be to prepare him for the fact that because he looks like a young Aryan he may hear things people don't intend him to hear, and he will have to unmask himself and defend his people. Once, a long time ago, a theatrical manager in Los Angeles suggested that since I didn't look Jewish, I should change my last name, a fairly common Jewish name, to something more neutral, maybe even the last name of someone already famous, to bring up more "pleasant connotations." As it was, he said, my name sounded like a girl people had gone to summer camp with. I told him, I probably *was* a girl people had gone to summer camp with, and as I remember it, we all had a damn good time.

*   *   *

This year, in a last-minute bid for a summer vacation, David and John and I decided to spend five days at a resort on the shore of Lake Champlain. We had never heard of the place, but it had been recommended by another place, which was already fully booked by people who actually *plan* vacations, and we figured we had nothing to lose but five days and a lot of money. It turned out to be a beautiful, family-oriented resort with tennis and golf and boating and a beach and a pool and activities for kids and even baby-sitting. The weather was glorious. Every morning we sent John to Kids Camp with a baby-sitter and we played tennis or kayaked or just lay there. Every afternoon we took John to the beach or pool and splashed and swam until four o'clock, at which time we broke for ice cream. Then we rode our bikes to the resort's grass airstrip, where we watched the little planes take off and land. "Plehn, plehhhhn," John shouted. It was perfect.

Well, *almost*. David and I both had this niggling feeling that something was missing. Something substantial. We just couldn't put a finger on what it was. And then, sitting in the dining room on the second night of our stay, right after the appetizer of selected crudités and just before the band struck up "The Girl from Ipanema," it hit us.

Jews.

There were no Jews. We looked around. Yes, three out of four men at the next table were, indeed, wearing kelly green blazers. A man on the dance floor sported plaid slacks. The women all had pageboy bobs and wore brightly colored shifts and shoes they had bought at Papagallo in 1975. We'd wandered into the Protestant version of the Catskills. There had been other clues. Only children and camp staff participated in the talent show. What Jew passes up a chance to perform in front of total strangers? Our round-robin tennis partners: Bets, Trish, Kath, Kev, and Chip (all right, it wasn't that bad but it was close) never swore at themselves or threw a racquet. Now, anyone who has ever played tennis at a Jewish country club knows that tournament protocol requires that every time you miss a shot you pound your racquet strings with the heel of your hand and bark at yourself: "Goddammit, Bobby, watch the goddamn ball." Also, the food stank.

I missed us, us Jews. It is such a comfort to have a few of us around. I was still glad we were there, though. It is our duty to maintain a presence in every corner of the globe. But at the same time I wouldn't be caught dead at the Concord or the Nevele. No Christians.

gone home

———◆———

I⊤ is gone. One moment it was there, as it had always been, utterly indispensable, and the next it was gone and forgotten. At some point, it seems, in the move from one wing of the nursing home to the other, The Pocketbook disappeared. My grandmother's lack of interest in this bit of news may seem unremarkable to most, to the uninitiated, but to those of us who know her it is indescribable. It is the end of an era. It is the beginning of the end.

The Pocketbook to which I refer was the last in what has been a long and illustrious line of pocketbooks. My grandmother was not a collector, nor were her pocketbooks particularly fine (except, of course, for the ones given her by my mother, the Queen of Quality), but they were like career soldiers, pressed into service year after year, decade after decade, their ribbons always straight and their shoes spit-polished. And unlike most pocketbooks, which maintain a fairly regular schedule of activities—marketing, Hadassah meetings, the Friday afternoon Philharmonic rehearsal (for members only)—The Pocketbook had special privileges. Indoor privileges. When my grandmother came to our house she didn't leave it in the foyer or in the guest bedroom or in any one place at all. It went with her from room to room and sat obediently, spaniel-like, beside her chair, at the ready. Glasses? Here. Tissues? Here. Small hard candies? Here. Wallet, checkbook, silk head scarf or folded plastic rain bonnet? Hail, hail, all here. Even in her own home, The Pocketbook made the circuit—bedroom, kitchen, living room, den. Other kids' grandmothers said things like "It's in my pocketbook, get it for me, sweetheart." Or queried, "Where are my mints?" Other kids' grandmothers carried satchels and straw bags big enough to transport a half-crocheted sweater, a pair of comfortable shoes, and four or five leftover onion rolls from the early-bird special at Rascal's. My grandmother only carried her Channel Thirteen canvas tote in the event of a sleepover or rain and never, ever to the exclusion of The Pocketbook.

No, a good leather pocketbook was essential. And she always wore good shoes and wool skirts and nice blouses and wool or cashmere (my mother) cardigans. She never wore sparkly sweatsuits or muumuus or caftans and she didn't own a pair of sneakers until I bought her some and they sat in the closet. I can't ever remember seeing her in a pair of slacks. Or saying the word *slacks*. Even now, somehow, she dresses as she always has, in her dangly earrings and her support stockings and her little sweater vests. When those go I don't know what I'll do. She was so unaffected by the loss of The Pocketbook, or rather, she didn't know it was gone, or rather again, she did not seem to remember it was ever there. I did everything but put out an APB. I went from desk to desk, nurse to nurse describing the thing: It's black, with a shoulder strap. You've seen them together a million times. Please help us.

Why did she let it out of her sight? I used to be inappropriately annoyed that she wanted to cart it everywhere; no stroll down the hall, no trip to the dining room was complete without it. I should have been relieved she was so unperturbed. But I was furious that it was gone. I don't need a therapist to tell me that I'm in denial, that I am resisting change no matter how inevitable, that she will eventually go the way of all pocketbooks. But goddammit anyway.

What is this process of letting go of things? It's really too painful to observe. A year or so before we moved her to the nursing home she began disengaging from her possessions. I don't mean as in "One day this cachepot will be yours." She wanted to give or even throw almost everything away. Now. Paintings, books, dishes. "What do I need it for?" became her mantra. If we'd emptied her apartment of everything but the teapot and her Estée Lauder bath powder she would have been thrilled. I'm not proud of this, but I actually became nervous she would promise things that I wanted one day, namely the piano, to whatever relatives happened to making their biannual pilgrimage to her door. In my head I was preparing speeches. "I'm sorry, I know she said you could have everything she owns, but you can't." I'm a big anticipator of confrontations that will probably never happen and this one repeated itself over and over in my head for months. I was attached to her things, well, some of her things, as I was attached to her, and I just didn't want anyone getting any funny ideas. One day my brother casually said, "Barbara and I are thinking we'll take the piano." "Think again," I said.

What was my grandmother going to do all day without The Pocketbook? It had provided her with any number of activities. She took things out, she put them in. If her glasses were missing or she hadn't brought along enough tissues, there was the task of going back to the room to retrieve the glasses or to the nurses' station to restock the tissues. Sometimes she carried around the small amount of jewelry we sent with her to the home. And there might be a comb, or some safety pins. How come all of a sudden these things were no longer essential to her? All right, so she lost The Pocketbook, but with it went the glasses for playing the piano and the pins for pinning whatever needed pinning. I became even more obsessed with what would occupy my grandmother's time and attention between our visits than I had been before.

Then, one day, I found her in the activity room, parked in her wheelchair among fifteen or so other residents in their wheelchairs, the big-screen television blaring in front of them—it was like a wheelchair parking lot, or a drive-in, really—and her head was bent and her hands were in her lap. I thought she might be asleep. But as I got closer I saw that she was fiddling with a piece of fabric. On another visit, as we sat in the winter garden, a large sunny room, she was stretching a corner of her skirt between her hands. On another visit still she continually smoothed a tissue on her lap. I finally realized what she was doing. My grandmother had been a dressmaker by trade and now she was fingering what little bit of fabric she could get ahold of as though she were examining it for a sewing project. She was planning where to put the pockets, how many inches to make the hem. After she stopped working she continued to make many of her own clothes and I used to bring her mine to alter. She had a black Singer sewing machine that flipped up out of its own maple table. Actually, it weighed a ton and had to be schlepped out with some effort. I would stand in her little den, turning slowly, one step at a time, as she knelt on the floor, measuring and pinning. I would sit with her while she basted, and then I would try the thing on again. All this before she even plugged the machine in. When she shopped for clothes at B. Altman's or Lord & Taylor, she stood in the store and turned out the sleeves of a jacket or examined the waistband of a skirt, determining the quality of the workmanship. One day, not long after B. Altman's had closed, my grandmother went to play bridge in her navy blue Altman's coat, and when her friends complimented the coat, without a word she opened it to reveal the Altman's label in the lining. "Aaah," sighed the women. Until the day we cleared my grandmother out of her apartment, those beautifully made suits and dresses hung in her closet, along with that navy coat, testaments to her glory days.

I was obsessed, as well, not just with what my grandmother was doing all day but with what she was thinking. Was she thinking "Make a skirt, make a skirt"? I hope so. That she may have been thinking "Where's Jack, where's Jack?" or Sam, her dead husband, or her mother or me or anyone, pained us all terribly. Or perhaps, by this time, her consciousness was devoid of the details; the body still knows how to do what it has done for three-quarters of a century, and the eyes see what they have seen a million times before. Sort of the phantom-limb theory. Before the Alzheimer's, when she was still just a wobbly old lady, my grandmother was afraid to hold her grandchildren in her arms. But later, some ancient maternal intuition reasserted itself and after John was born she reached for him. Even though she was hard pressed to remember that she herself had a son, her body knew what it knew. And "I know you, I know you," is exactly what she would say to the baby, each time she held him. It was one of the few coherent phrases she had left. Sometimes I thought that as she lost language and John gained it, there would be a meeting place, and they would be able to communicate only with each other.

But I need to say this: She never didn't know me. Never.

I think about the person my grandmother once was, or rather, how I will want to remember her, one day, some days from now. She was brave and resourceful. She practically raised her sisters and once she painted her living room in Long Beach in a single morning. She loved her family fiercely. She was not daunted by her inability to play bridge well. She was not boastful about her gifts as a classical pianist. She never flinched from life's burdensome tasks. She never stopped trying to better herself; who else would slog through Thomas Pynchon at ninety years old? She volunteered for ORT and Hadassah. She was a pioneer, really. She was fourteen years old when she arrived in New York with her mother and two of her three sisters. They had worked their way across the continent to earn their passage, from Odessa to a port in Morocco, maybe Tangier or Casablanca. Of her childhood in Russia, I knew only that her father, from whom she had been inseparable, had died young and poor; that the pogroms had begun; and that one day she stood in her school uniform with hundreds of other children to watch the Red Army march through the streets of Odessa.

In Grafton, Vermont, not far from where my husband's family has a vacation house, there is a cemetery. Unless you know it is there it is easy to miss. It is at the very edge of the town center, through a rusty turnstile and up a steep path. In the summer small bees hum drunkenly over rampant wild thyme, and in the winter it is foot deep in snow, sheltered from the wind by towering, white-tipped pines and naked gleaming birches. There are two aisles, one directly up the center and the other transecting it about three-quarters of the way, and when you stand in the back of the Grafton Cemetery you know exactly where you are: in Church. (I know, I'm a Jew, but just go with me.) Most of the gravestones face forward up the nave in neat rows and the rest, the most prominent, face back at them across the front aisle. It is as though they are where they have always been, where they belonged, before anything bad ever happened, before illness and war and lumber accidents. It also goes to show you that when you take the big plane ride to eternity, your seat assignment doesn't change, which seems both a comfort and an outrage. Still, this is an awesome, soul-piercing place, inhabited by two centuries of New Englanders past, their epitaphs fading stoically from their stony faces.

There is one that I will always remember: GONE HOME.

Maybe The Pocketbook has done that. Maybe it has gone back to the apartment, looking for a familiar face—the den chair, the kitchen table, the piano. Or back even farther, to Brooklyn or Philadelphia or Odessa, as my grandmother does when she asks for her mother and sisters and husband as if they were all still alive. Maybe The Pocketbook didn't recognize my grandmother anymore and went in search of a youthful-looking senior playing bridge at the Ninety-second Street Y or shopping in Altman's for Ferragamos. Maybe it couldn't get used to her gradual decline, as I have, and can't shake the memory of her old self, as I usually can. I have said over and over to people, as if trying to reassure them, or myself, that I don't really remember my grandmother before the big slide began, that it has come over us both gradually and in a way that has continually displaced the past with its own overpowering present. But now, writing this, I remember her helping me turn my old jeans into a skirt on the sewing machine she gave me for my thirteenth birthday. I remember the day she came to see me in a play wearing two different blue shoes, and how we laughed. I remember eating potato chips with her in the hospital cafeteria while my grandfather was dying upstairs.

All my life she never seemed to age. I grew up and she just sort of hung around, as is. But now, suddenly, she is blasting forward and I feel stuck. I am the adult and she is the child and she has lost her pocketbook and doesn't care, doesn't see what it means, that things have ramifications. But, of course, they don't. Not for her. Only for me. Some day, I will have an actual conversation like that with my child. I will say: "Don't lose the things you need." He will answer quite logically: "I don't need them anymore."

the good swimmer or
how I lost him

———

LIKE a swimmer approaching Dover from Calais, my husband is slowly looming into view. I can see his capped head, bobbing up and down in the distance, rising and falling between the frothy white ridges, disappearing, reappearing; yes, he's coming. I can see him now with the naked eye.

When my son emerged from his watery nest in my belly last year, my husband, conversely, submerged. He dissolved with a celebratory fizz followed by silence, like the flattening of champagne, like a bromide, into the murky depths of fatherhood, which is nothing akin to husbandhood. It was not at all his fault, but my own. I cast him off. That I did so unwittingly is not much of an excuse. But I did not know that my son would displace him so thoroughly nor that my episiotomy would hurt for almost a year. So I agreed that when the year was up we would go to the wedding of a friend in the Bahamas without the baby. I agreed to this many months in advance, under the delusion that by the time the wedding rolled around it would be good to have some time alone with my husband, some adult time. What this meant, I could only vaguely recall. They don't tell you until you're too deep into it, but breast-feeding really kills the libido. You know, I hate that word, *libido*. It's so ugly. It would be better backward—odibil. Breast-feeding, and exhaustion, and fierce, all-consuming mother love, are no friend to odibil. And yet my attraction to my *son*, who has staked out the territory between my collarbone and my belly as his own, and who has, as well, mastered the art of French kissing, is terrifying. It is the love that dare not speak its name. Well, not really, that's something else. But the kid sure can kiss.

Still, when you have to go to the Bahamas, you have to go to the Bahamas.

Okay, we had a decent time, so sue me. It was nice to sleep late and read a whole book (on the beach!) and eat a meal in more than five minutes and have painful sex without the Baby of Damocles hanging over our heads. It was fun to have twenty-four-hour butler service, although try as we might, we couldn't come up with anything that needed butling. And we were only a fifteen-minute walk down the beach from the Atlantis Resort, a thirty-trillion-guest-room colossus intended to conjure the mythic sunken city. It didn't.

I'm no authority on the subject of mythology, but I don't recall that the denizens of that great underwater metropolis at any time engaged in poolside aerobic and spinning classes. Nevertheless, the Atlantis Resort did have a spectacular network of water slides, and we and other wedding guests frolicked like seals on crack. When we'd had enough of the fleshy, sunstroked throng, we returned to our gracious little hotel for lunch.

Throughout the weekend friends and acquaintances inquired after John, and if by day two my response was "John who?" it was definitely meant as a joke.

Because on the way home all David and I could talk about was seeing the kid. The excitement was enormous. I foresaw a joyful homecoming with smiles and hugs and a lot of filthy, disgusting, Oedipal behavior. When we walked in the door there was a mad dash for him, but I elbowed David in the kidneys and got there first. I held John for a long time. I didn't notice that he was not holding me back. Not in that way he has; usually he clings like a monkey. He seemed alternately bemused and insouciant, which was understandable, although some trick for a baby to pull off. Suddenly his parents show up after a three-day absence, just when he'd gotten used to his grandparents being his new parents; what the fuck? But I was so happy to be with him I just didn't notice. I was naive.

The next day the baby avoided me entirely. He wouldn't even look me in the eye. Aloof is putting it kindly. He turned all his attention to David, as if David hadn't gone away, too. As if David hadn't disappeared for four days on some idiotic boondoggle just a month earlier. Or gone to China for eight days in the spring. Although John had only been walking for a few weeks, he navigated around me with surprising finesse. He started toward me and then, *psych,* turned away. Toward David. Toward anything. Suddenly the Word was *Da-dee.* "We're not British, you know," I said to him, but he even ignored the insult.

And what I immediately realized was this: I blew it. I blew the best, most important relationship I have ever had, with the only person who has ever loved me completely, fanatically, and certainly unconditionally. I waited until he was just old enough to be cognizant of my whereabouts and then I jetted off to parts unknown without a word of warning. As far as he was concerned, we were through.

It's not like I didn't expect some sort of reaction. But he'd had such a good time with my parents, or so they said. He'd been no trouble at all and hadn't been disturbed in the least when he woke up in the morning to find that despite all efforts to the contrary, his mother really had turned into her mother, after all. What I had thought would happen was the following: he would cry often, because I was not with him, but he would persevere, and then, when I came home, he would be ecstatically relieved and cleave to me like a deer tick. But that is not what happened at all. Instead he bided his time, scheming and plotting, so when I finally showed up he could exact his cruel punishment.

Over the next two weeks leading up to Thanksgiving, I did everything I could to win him back. I dressed him in his favorite clothes and made him his favorite foods. We played with his favorite toys and sang his favorite songs. I figured out the guitar chords to "John Jacob Jinglehiemer Schmidt." I showered him with praise. He remained indifferent. He suffered me. He adored David. I hated David. How many children spend their lives trying to please fathers who come home at seven o'clock each night and play golf all weekend? Fathers who somehow manage to arrive on the scene when all the barfing is done. Actually, David is a much better father than this, he has to be or I'd divorce him, but still, fuckity fuck fuck, I gave that baby life, the ingrate.

I was never popular. I wasn't exactly unpopular, I mean I wasn't beneath the radar; I was somewhere in between. In high school I had a group of close friends and we sort of rummaged around on the outskirts of the popular crowd. We were welcome at parties but not essential to anyone else's good time. I was part of a gang in college, I guess, but I never got the impression that when I wasn't there someone said, "Hey, where the hell's Kaplan?" I still dream sometimes that my college friends are planning something fun and don't include me. Sometime around my twenty-seventh year, I was really, really popular for a week.

Here's what it is: I was *extremely* popular with John. I was at the apex of my popularity potential. I *was* the in crowd as far as he was concerned, the prettiest girl in class, the highest scorer on the team. When I wasn't around this kid was definitely thinking, "Hey, where the hell's Mommy?" Wherever I was was the cool place to be. This concept is at the very heart of popularity. A cool kid could stand in a pile of shit and people would say, "Hey, Bob's in shit, let's go!" Of course, you could say that it was really *John* who was popular with *me*. He often did stand in shit and I was right there by his side every time. But in fact it was our *mutual* regard that made us what we were, that makes all great friendships what they are. One popular person is a magnet, two are a fortress. People came along with John and me, not we with them.

And one of those people was David. Because where John and I were the most popular was in our apartment.

You see, it doesn't matter if you are popular in a city of one million or a class of 150 or an apartment of three. As a matter of fact, I've been in threesomes (not that kind) before, and there has always come a moment when the odd man, i.e. me, is out. Well, not this time, friends. Not this time.

But then I blew it.

I opened the door and let that annoying hanger-on, David, in. He'd been there, swimming around the island, treading water, blowing bubbles, whatever. And somehow I thought that he was waiting for *me*. For me to come around. To loosen up the tea party. To make room for him. For sex. Instead, though, because I'd left John alone on the beach like a baby turtle, so to speak, and only for a *second*, I tell you, David swam up to shore, opened his big fat yapper, and took the kid whole.

horse kills owner or
how I got him back

———

THE day after Thanksgiving, David, John, and I drove to Vermont, hoping to do a little skiing in the preseason lull, but by the time we got to the ski house John had a fever, and we canceled our reservation at the mountain's day-care center for the next day. In the morning, I sent David skiing and I stayed with John, coaxing him to eat a few Cheerios and sip water. I fed him drops of Motrin and when the fever dipped we sat on the orange shag carpet and put the colored balls into the translucent tube and then took them out again. David came back early and I went to the mountain and skied for an hour. When I got home John was slumped on David's shoulder, which is where he spent the rest of the day. That night, he woke up crying, very hot, and we took him into our bed, where he slept curled against my body, ripples of heat rising from his skin, bending the air.

By early Sunday morning his fever spiked to 104 and he was alternately lethargic and howling, so we took him to a doctor in Manchester, who, after examining him, sent us to the hospital forty miles away in Bennington.

We signed in with an admitting nurse, who took John's temperature and gave him more Motrin, and then we took seats in a waiting area that wasn't much bigger than the reception room of a large dental practice. There were only two other people there, and the nurse told us we could expect to wait about an hour. An hour went by and the nurse came out and said that there had been a serious accident and that we might have to wait longer. Unable to help myself, I asked if it was a car crash—the weather was terrible, cold and rain and fog—but she said, "No, a horse." A *horse*? David took John for a walk in search of ice cream, and I picked up an old copy of *In Style* magazine and turned the pages without looking at them. A horse.

A horse accident was exactly what I needed. One of the things about being in a hospital is that in order to cope with your own medical crisis, you become obsessed with the medical crises of total strangers. You have to know every diagnosis, prognosis, and outcome that occurs within the constantly shifting thirty-foot radius of your own earshot. Did I say *you*? I always say *you*, but what I really mean is *I*. *I* have to know every diagnosis, prognosis, and outcome that occurs within the constantly shifting thirty-foot radius of *my* own earshot. Just my proximity to the surrounding real-life drama gives me some kind of lurid license to shoehorn myself into it. *I was there when . . . I saw . . .* What? What did I see? Well, while I was waiting for David and John to come back with ice cream, I saw myself as part of the cast of *The Shooting Party,* a movie starring James Mason in which a bunch of romantically entangled people are staying in a manor house for the weekend and a bird-shooting expedition goes terribly wrong. There is riding in the movie, and Englishmen. It doesn't matter at all that in real life I don't like horses and hate riding. What matters is that it was easier to think about an accident with a horse than to let myself imagine any one of several terrible potential scenarios starring John.

There are many times, particularly in the first year of your first child's life, when you are absolutely convinced he has the Bubonic Plague. And if not the Bubonic Plague, then the Spanish Flu or Mad Cow Disease or that disease you get from being in a barn with owl droppings. Every fever is terrifying, every cold ominous, and you rush to the pediatrician's office so he or she can proclaim, "It's just a virus!" Like "It's just a paper cut!" You feel ridiculous every time you take your baby in. You are a time-sucking alarmist, clogging the works with your piddly worries while truly sick babies languish at home for want of an available appointment. (I *am* a time-sucking alarmist, but that's not the point.) Fortunately, with experience, your paranoia settles into your subconscious in a way that will sometimes keep you up at night but will most likely allow you to behave rationally during the day. That is, until the day a doctor sends your child to the hospital. In a way, it is as if you have been waiting all year for this moment on the condition, of course, that this moment will never come.

David and John came back and we were ushered into the triage area and assigned an examination room. A nurse arrived to take John's blood, but she had trouble finding a vein. She called for another nurse, and while David and I held John down, they tried, and failed, to draw his blood. His veins were deflated from dehydration. Horrified, I imagined John's blood hardened into a dry, crystalline substance in his veins, like Pixie Sticks. The nurses pointed out that despite the fact that John was crying, his eyes were dry. "No tears, that's a sign," they said, and they left us with a cup of ice chips.

Then a technician came by to take us to X ray. In order to X-ray John's chest, the technician insisted (I want it on record that we argued with her) on strapping him into a contraption that could only have been descended from a medieval torture device, like the rack, and was improbably called a Pigg-O-Stat. I could not even guess to what the name referred, except the obvious, and I have never heard of a pig getting a chest X ray, never. All I could think of was those squealing, writhing pigs that some people, somewhere in this great country of ours, grease up and then try to pin down, for fun and/or a blue ribbon. Honestly, couldn't someone in marketing have come up with a less disturbing name? The Pigg-O-Stat was made mostly of Plexiglas, an outdated material if there ever was one, and it consisted of a little bicycle seat, upon which the unsuspecting baby was perched upright, and a back and chest plate, also Plexiglas, which came together with leather straps and immured its tiny victim like a sausage casing. *Saucisson de bébé.* David and I stood, shocked, behind a heavy door, while the technician clicked away and John wailed in fright. When it was over, she took a moment to review the X ray. She put it in a light box and made a face like Whoo! and then refused to say anything to us. We returned to our holding pen and soon a resident came in and announced that John had a big, whopping pneumonia.

Then the pediatrician arrived from her home five minutes away and expertly inserted an IV line in the back of John's hand, and we were taken upstairs where we stayed for two days and nights.

Bennington is not a very big city. I'm not sure it is a city. It might be just a town. Or a hamlet. At its center is Bennington College, a liberal-arts school where I have always imagined modern dancers are continually performing on the lawn, whatever the weather, and which my bookish mother attended. Twenty minutes south is Williamstown, the home of Williams College. When we pass by Williams during the school year on our way to the ski house, our car slows to a crawl and David and I both strain to see into the lit dorm rooms, to relive for a millisecond our college experiences as they did not happen. Sometimes I think about how I never got to audition for the Williamstown Theater Festival and probably never will. The border between Williamstown and Bennington is also the Massachusetts–Vermont border, and as we cross from one to the other, there is a perceptible shift, the countryside yawns and opens, the light changes. We drive north past dairy farms and two-hundred-year-old houses and corn fields and hay fields, the configurations of which have hardly changed for centuries. It is starkly romantic, at once inviting and fearful, like *Ethan Frome* (except there's a little more money to go around and three-quarters of the way through, Zeena, thank God, dies). I stare silently from the car window at the pale dusting of snow and the black trees and imagine, you know, stuff. That I live in a different time, when things were, um, different. More. More something. More horse accidents. Not everyone makes it back from the hunt.

I am not one of these New Englanders who marvels at the glory of autumn. Sure, show me a flaming red maple and I'll say, "Wow," but really, I am just biding my time until the rot sets in. I am waiting for the leaves to curl in upon themselves and fall and become brittle beneath my feet. I am waiting for the pumpkin pickers to clear the pumpkin patch and leave the last, lonely, misshapen pumpkins to their soggy fates. I am waiting for the end of the harvest so I can get a clear view of those empty brown rows, the dry stalks, the bare branches. To me, late fall days carry an electrically charged current of dread. The winter is coming, maybe it will snow, something could die. It is not that I *want* something to die, but I am peculiarly energized by the idea that it could.

All of my life I have romanticized death. I do it in a vague, movie way, something like: my husband dies suddenly (not actually my real husband, just some movie husband, the one I live with in the Movie of Me, which runs on an endless loop in my head) and I am thrust into the center of attention. That's what a death does, for someone who has never experienced a death, that is. It makes you the center of attention. People gather around you and protect you and shush everyone else and you become a tragic movie figure and are eventually courted by a man who has loved you from afar for years but you always thought of as a friendly acquaintance until now. More likely than not he has an English accent. The Movie of Me is all about redemption, about reawakening. It is about starting over and being newborn and shaking off my old self: Now that I am who I am, maybe I can start again and not be who I was.

I worry that this kind of fantasizing is dangerous (puh, puh, puh), although, God knows, I've dreamt often enough that Tony Kushner decided to start writing plays for me to star in and *that* hasn't come to pass. And, of course, all this flies in the face of my almost pathological fearfulness. How is it possible to live in an constant state of dread and simultaneously court it? Easy. The same deviant impulse that tortures me with insomniac visions of pain and loss redeems itself with the dream-promise that when it is all over I will have amazing sex on the castle floor.

The hospital provided twin beds for David and me. I felt like we were on some kind of hospital dream vacation. In New York we would have been lucky if one of us had been able to sit up all night in a chair. Unfortunately, John was not as appreciative of his accommodation—a metal cage—so we pushed the beds together and David slept in one bed, I slept on the crack, and John slept next to me in the other bed. Every hour throughout the night, someone in pastel scrubs checked his temperature and his blood pressure, replaced an empty saline bag with a full one, adjusted his antibiotic drip, or wafted medicinal fumes across his face. John cried out and tossed, and I stroked his forehead and whispered to him and rearranged his IV lines around him. Between these ministrations, I lay awake, speculating in agony what would have happened if we hadn't brought him to the hospital, if we'd tried to ride it out, to "see how things go," in the words of the pediatrician. How the hell had we missed the tear thing? That baby had cried and cried, yet his eyes were dry, were we blind? Children all over the Third World die of dehydration daily. We have plenty of fluids at our disposal, clean ones, flavored ones. Of course, if I'd still been *nursing*, he'd have been belly up to the bar. But no, *I* had to go to the Bahamas.

And he'd been congested all week, coughing each morning like a tubercular old man. And he'd thrown up on Thanksgiving. Actually, I had gotten myself into a bit of a tizzy over the throwing up, but my entire family practically chanted in unison: "Kids throw up. Kids throw up. Kids throw up." I hated them.

The night wore on. It is impossible to sleep when you are angry at yourself; all you want to do is punch yourself in the head. I finally nodded off by deflecting my anger onto my brother, who never put a gate at the top of the stairs when his girls were small.

By the morning John's fever had ebbed, and he drank juice and ate a little and wandered around our room and the adjoining playroom (the place was bigger than our apartment), one arm opening and shutting the door of the play oven, the other arm, a comical mummy arm, wrapped to the elbow in gauze and covered with a diaper to protect the IV line. David and I trailed him, rolling his IV pole, relief washing over us in endless waves.

At exactly this point, I began to think of myself as The Mother in a Lifetime Network "Moment of Truth" movie: *For the Love of Her Child: The Cynthia Kaplan Story.* There was a little kitchen on the pediatric ward and every time I went to fetch some milk or a box of raisins or Cheerios, I imagined the camera following me: *There she is, the woman whose baby is sick.*

I also felt compelled to find out what had really happened

with the horse, so I asked around the ward—discreetly, I thought, although David thought otherwise—but no one had heard about it. As we prepared to leave the hospital on Tuesday afternoon I bought the local paper, the *Bennington Banner,* out of a coin-operated box in the lobby. ("Serving the communities of Bennington County." County!) I scanned the front page over John's head as he sat in my lap. There it was. HORSE KILLS ITS OWNER IN ACCIDENT. A forty-six-year-old woman, recently transplanted from "down-state" New York, was preparing to ride the horse she had owned for fifteen years, a gentle horse, by the account in the paper, when it broke away and ran out of the barn and into a field. As the woman and her boyfriend tried to regain control of the horse, it kicked her in the chest. The boyfriend carried her in from the field, screaming for help.

Oh. Nothing is ever how you think it is going to be. Or how you imagine it should be.

As we drove away, I stared out at the barns and the fields and the gray Vermont sky. They did not look as they had. I cried then for the first time since all this had begun. I kissed John's sweet, cool cheek, and then I closed my eyes and let David take us home.

the few, the proud

———

WHAT's the story with those French truffle pigs? If they like truffles so much how come they don't just eat them? What's stopping them? What's stopping them from just saying to those French truffle farmers, "Buzz off, *monsieur*, I saw it first," and then snarfing it down? I'll tell you what's stopping them. Muzzles and leashes and whaps on the snout with a knobby walking stick. *That's* what's stopping them. Jesus, how would you like it to be your portion in life to constantly be searching for the yummiest thing you can think of, the thing you want the most, and then every time you find it someone schleps you back with a jerk, snapping your head probably and whapping you on the nose for good measure, just to teach you a lesson, and *then* takes it for himself. Just takes that yummy thing. Just takes your happiness. And in the bargain, *in the bargain,* serves you slop for dinner. If you were smart, you truffle pigs, you would suck those truffles right up and then act cool, like that one wasn't a truffle at all, whoops, you made a mistake, that was just a regular little brown mushroom, not a truffle, no, definitely not a truffle, yech. But you're not smart, you're just a fat stupid pig and the only thing standing between you as you are now and you as a Jimmy Dean Pork Sausage is your keen sense of smell. And people pat you on the back or scratch you between the ears and tell you what a good little pig you are, you're so talented, so special. And you let yourself believe it, for just one moment, one golden shining moment. I'm special, I'm talented, I have an Extraordinary Sense of Smell. And that's how it starts. You get sucked into the vortex. No, no, you go willingly, you weak, easily manipulated piece of shit. Hi, I have only one discernable talent, no ambition, and no self-esteem. May I please go into the vortex? But don't feel badly, because it's not really your fault. You have to live, for God's sake, you have to eat and sleep and work just like everyone else. So you compromise, you give a little, and you give a little more and a little more until you're just a pathetic tool, a loser, an undignified, spineless . . . pig.

And you dream of freedom. You dream of the day when you will run muzzleless, leashless, through the truffled woods of your youth, full of promise, full of hope. You dream of release, of deliverance, from your hell, from your prison, yes, the one you built, from your own personal Church of Scientology. But *is* there any way out? *Is* there? You ask the other pigs. But what do they know? They're just pigs. Oh, God, let me out! Let me have my happiness. Let me eat my truffles! Let me eat my fucking truffles! Or not. Not if I don't feel like it. Maybe I'm in the mood for a truffle. Maybe I'm not. It's my choice. I have a free will. This is America, goddamn it. Oh, wait, no, shit, it's France.

But eventually you make your peace. Nobody has a perfect life, you say. Nobody gets everything. You have your health, for God's sake. Things could be worse. Much worse. Some pigs have never experienced the remarkable, earthy intensity of the truffle. Some pigs have never been delirious with the heady fragrance, or dizzy with the thrill of the find. I found it, I found it! Some pigs wouldn't know a truffle from a stale Devil Dog—which, as *you* would know, because you are a truffle pig, smell almost exactly alike.

And you'll tell your children and your children's children that *yes*, you were part of that elite society. You were one of the few, the proud. You, a truffle pig. How many can claim that distinction? And finally, at the end of the day, after the reminiscences and the songs, after the anecdotes, after *all*, you will have convinced yourself, as well. You will have convinced yourself that you did your best, that these were the chips you were dealt, this was the life you were destined to live. Because to think otherwise, to think otherwise, my friends, would be unbearable.

Acknowledgments

In acknowledging the following people for their various and sundry contributions to the writing of this book, I am using a sequencing method that I prefer not to divulge, thus leaving people to wonder why they were thanked before or after a certain other person. If I have left anyone out, it is because you were of no help to me whatsoever. If I have named someone twice, in italics, or in bold, it is because I love you more and thank you more than the others.

Neil Genzlinger, the writer and editor, who put me in print and befriended me to boot; Michael Murphy and Doris Cooper, who generously invited me to write a book; Marjorie Braman, my *very* wonderful editor; Amy Rennert, my brave, smart agent; the great Lisa Gallagher, the entire Morrow sales staff, and the incomparable Dee Dee DeBartlo; Jill Lamar, whose support changed everything; Nancy Giles, Steve Olson and Hank Meiman, Mason Pettit and Gregory Wolfe et al., Victoria Rowan, Ellie Covan, John McCormack, Liz Benjamin, Joe Danisi, and Naked Angels, for years of support; audiences of: *The New Jack Paar Show, Moonwork, Stories on Stage, Tuesdays at Nine*; Amy Krouse Rosenthal, a comrade even though she once dated my husband;

Amanda Green, the smartest reader, the best friend; Rholda Hyacinth, trusty babysitter; my Family at Large—all Kaplans, Froelichs, Logans, Mosses, Schwartz-Halls, and Siegels, for their bottomless support; my brother, Steven Kaplan, for my taste in music and for making sure there was a witness; my grandparents in absentia: Benjamin and Lillian Siegel, Sam and Dorothy Kaplan—how they loved me; my remarkable and forbearing parents, Sandra and Jack Kaplan— how I love them; John Benjamin Froelich, my beautiful child, without whom this book would be very short; David Froelich, my beautiful husband, without whom there would be no book.

About the author

About the book

Insights,
Interviews
& More . . .

Read on

A Conversation with Cynthia Kaplan

© Bill Westmoreland

Salt Magazine's Terra McVoy subjected Cynthia Kaplan to one of Salt's "Writer in a Jar" interviews. It did not last long and was not painful and when it was done they let her out of the jar. It is hereby reported with permission.

Terra McVoy: *Any writers who have influenced you?*

Cynthia Kaplan: James Thurber. Early Woody Allen. Edith Wharton, Pete Dexter, Alice Munro. I realize the last three are not known for their senses of humor, but they write with extreme honesty and no filler. I have read *The House of Mirth* about thirty times and there isn't an unnecessary word. Also, an American writer named Rachel Ingalls, who lives in London and wrote two of my favorite books that no one has heard of but should have, *Mrs. Caliban* and *Binstead's Safari*.

T.M.: *What, if anything, are you reading now?*

C.K.: Unfortunately, I've developed an unhelpful habit of reading the *Post*'s gossip columnists online while I eat lunch. I read a lot memoirs and essays, sort of as research, to see what the opposition is up to.

T.M.: *Three of your favorite Olympic competitions?*

C.K.: Slalom, Giant Slalom, and Super-G. I once wrote a piece about introducing a painting competition. I'd like to see this happen.

T.M.: *How heavily do you rely on research for your work?*

C.K.: Most of my research involves calling my mother and asking her the name of some obscure person or event or movie that only she would remember. Sometimes I use Google, but I'd call that procrastination rather than research.

T.M.: *Talk about your favorite/least favorite rejection.*

C.K.: My favorite rejection was from a literary agent who in her letter called me a very talented writer, but spelled talented with two L's.

T.M.: *Any particular writing habits you have? (Music you listen to, where you work, etc.)*

C.K.: I would do well to develop some writing habits. The ones I have now involve sitting at my desk and trying to write. I try to write every day. At least I sit in front of the computer every day. I can only write when I know what I want to say. It comes in bursts and I can have a very productive hour or two in the middle of a lot of staring into space. I also have to work around my pesky children, who like attention.

T.M.: *Tissues vs. handkerchiefs: any comments?*

C.K.: Ugh, tissues, of course. Handkerchiefs are just not absorbent enough. And you have to keep sticking the thing back in your pocket or up your sleeve or in your purse or whatever. After I blow my nose I never want to see whatever came out of it again.

T.M.: *What do you do when you get stuck in your writing?* ▶

A Conversation with Cynthia Kaplan *(continued)*

C.K.: Snack. Pee. Check my Amazon rating. Check other people's Amazon ratings. Read other people's writing. If I read something great, I feel inspired; if I read something terrible, I feel superior, and that helps me write.

T.M.: *Can you talk a little bit about your process? Meaning, how your work develops from idea to actual finished product?*

C.K.: I get an idea in my head. Maybe it is a memory, or a phrase, or an idea about something I saw. I try to write it down somewhere before I forget it. Then I think about why it struck me in the first place. Or sometimes I'll start riffing on a subject at dinner or in the car and realize I have an opinion that I want to explore. And then maybe I'll realize that opinion connects to an experience I've had and I'll slowly work to tie them together.

Sometimes an essay pops out in a day and sometimes it takes months. Sometimes I write a paragraph and don't get back to it for six months because I don't know where it is going exactly. When I know, then I can write. I always, always have to have determined my point of view. For me, that's almost everything. I don't have weird and wild stories to tell, I only have my voice.

I also self-edit furiously. I'm easily bored, and if I'm bored, I assume the reader will be, too.

T.M.: *Most irrational fear you have?*

C.K.: Huh, where to start? Moths.

When I lived on the thirty-third floor of a building, it was having a plane crash into my apartment. Unfortunately, that stopped being so irrational.

T.M.: *Do reviews influence you at all? How important are they to you and motivating your work?*

C.K.: Good reviews don't motivate my work, but they sure make me feel good. Bad reviews don't bother me that much because I usually think they're wrong.

T.M.: *When did you know you wanted to be a writer/performer?*

C.K.: I knew I wanted to be an actor from a very young age, maybe like three months. In school I used to say I wanted to be a writer so that I didn't have to say I wanted to be an actor, which sounded insane. But then, at some point, I actually started writing, and then eventually I found I sort of HAD to write.

T.M.: *How did you get your start?*

C.K.: I started writing comedy in my twenties after taking a week-long workshop with the comedian Lewis Black. At the end of the week he encouraged me to do it in the real world, as opposed to the excellent, insular summer workshop world. A few years later, after a lot of writing and performing, a friend recommended me to an editor on the Op-Ed page of the *Times*. I sent him some monologue/essay sort of things and he started giving me assignments to write humor pieces for the Op-Ed page. One of those pieces led to a book deal.

T.M.: *If you could have any other job, what would it be?*

C.K.: Legitimate investigative journalist whose hard-hitting reports reveal the corrupt inner workings of big business and our government. I'd also like to be a successful stage actress. I guess these aren't really other jobs, just more glamorous versions of the jobs I already have.

T.M.: *And what is your favorite thing to do on a rainy Thursday evening?*

C.K.: I'm not sure how to answer that. I can't figure out what the key element is. Rain? Evening? Thursday?

T.M.: *What are some of the biggest distractions for you in terms of your work?*

C.K.: My children and my disastrous personality.

T.M.: *How important do you think it is for writers/performers to live in an urban area?* ▶

5

A Conversation with Cynthia Kaplan *(continued)*

C.K.: Depends. If you are a performer, you can probably scare up a bigger audience for your performance if there are more than ten people per square mile living in your town. If you are a writer, I guess, you should live wherever you write best, wherever you feel most like yourself, or the self you want to be when you write.

T.M.: *Any thoughts on big industry vs. independent publishing?*

C.K.: No thoughts at the moment. By the time I have one there will probably only be one publishing company in the world, publishing books about cats and the wretched Bush administration, which will be its only saving grace.

T.M.: *What is the hardest part about being a writer?*

C.K.: Being a writer.

T.M.: *The most rewarding part?*

C.K.: Being a writer.
　　Sorry. Those last two questions were very James Lipton.

T.M.: *And lastly . . . Matt Lauer: Hot or not?*

C.K.: I prefer Chris Matthews.
　　Actually, I really prefer Eddie Vedder. ∾

Truth Be Told

WHOEVER SAID Truth is stranger than fiction *didn't know me.*

People often ask me if my stories are true. I am surprised by this question because in my opinion nothing that happens in my book is so extraordinary that it would be hard to believe. I didn't know how to slow dance, my therapist went nuts, my father likes gadgets. Why would anyone make up these things? They'd make up something better. Anyway, for you doubters out there, I offer an example of what an untrue story by me looks like. I wrote it after the movie Face/Off *came out, in the hopes that* The New Yorker *would publish it as a* Shouts & Murmurs. *They didn't. Wonder why.*

Brain/Out

YESTERDAY MORNING at exactly 5:57 a.m., my husband David exchanged brains with our basset hound, Slowpoke, in a pain-free, non-scarring operation, in the hopes that he would be able to convince the neighbor's bichon frise, Charles of the Ritz, to reveal the whereabouts of our garage door opener. Garage door openers, as we all know, are expensive to replace and the last time we asked our neighbors to retrieve something of ours from their yard the husband threatened to come over and beat the b'Jesus out of us. Seeing as both my husband and I are of the Jewish persuasion, that probably wasn't going to be very easy. It could take a lot of time and many punches about the head. We felt that under the circumstances the surgery was our only option. ▶

> 66 Whoever said Truth is stranger than fiction *didn't know me.* 99

7

Truth Be Told *(continued)*

At 7:30 I picked up David and Slowpoke from The Clinic. David sat with his head out the front passenger window, his tongue hanging out, and yelled obscenities at rocks, and Slowpoke circled twenty-three times around a picture of George W. in the *New York Times* and then peed on it. They could have been their old selves! Let's hope Charles of the Ritz thinks so, anyway.

When we arrived at the house David leaped from the car and disappeared into the woods. I was worried about him roaming the neighborhood, sniffing between people's legs and whatnot, but first things first, and that meant the garage door opener. I took a magazine and Slowpoke took a squeaky toy and we lounged casually in the back yard waiting for Charles of the Ritz to scamper over, which he invariably did. Slowpoke and I pretended not to be watching while Charles of the Ritz made off with the squeaky toy. Then Slowpoke followed him with a stealth worthy of a dachshund. His sudden grace reminded me of David when we were first dating. I wanted to call him back, but, well, as I said, first things first.

David himself returned to the house around dinner time with some dead leaves in his hair and only one shoe. He ate some kibble and a half a Gainesburger and went to sleep in front of the TV. Slowpoke wasn't back yet so I warmed up last night's risotto and watched an old Bette Davis movie, *Dead Ringer*, about a pair of murderous twins who trade identities. Finally, I heard a scratching at the back door. I could tell by Slowpoke's sad eyes and droopy face that he'd had no luck. We headed upstairs without speaking. What was there to say?

66 When we arrived at the house [my husband, David,] leaped from the car and disappeared into the woods. I was worried about him roaming the neighborhood, sniffing between people's legs and whatnot. 99

Sometime later David came and lay down on the end of the bed. Slowpoke held me tenderly. We slept.

The next day Slowpoke was out early, trailing Charles of the Ritz around the neighborhood. David was God knows where. I waited. Then, suddenly, around 3:30 in the afternoon, all hell broke loose. I raced through the house and out the back door and there were Charles of the Ritz and his owners all barking and screaming at the tops of their lungs. Slowpoke was running around in circles and at the center of the mayhem stood David, covered in dirt, the garage door opener in his mouth. Inexplicably, this final scene took place in slow motion.

At 5:34 a.m. this morning David and Slowpoke were restored to their former selves and I was grateful. Grateful that I do not have to lie about what happened between myself and Slowpoke because nothing did happen, and grateful to have our garage door opener back. ◠

❝ At the center of the mayhem stood David, covered in dirt, the garage door opener in his mouth. ❞

Reading and Eating

ONE OF LIFE'S GREAT PLEASURES is reading while eating. Or eating while reading. Below is a list of foods I would recommend you eat while reading my essays. If you have already read my essays, read them again, because eating may very well enhance the experience. In fact, it will probably work best if you buy yourself a second copy of the book, so you don't get food all over the first copy. Or else buy my next book and eat while you read that. Or read *Moby-Dick*, whatever, it's your life. Anyway, here goes:

queechy girls

The obvious suggestion would be s'mores, but they are too labor-intensive and too messy and most campfires don't produce enough light to read by and books are flammable. Rather, I recommend a small bowl of Nesquik powder, or if you're a fruit drink kind of person, presweetened Strawberry Kool-Aid mix. Lick your finger, dip, repeat.

a dog loves a bone

Mini hard salami links. Peal the plastic casing off ahead of time and cut into small, quarter-inch-thick rounds. Finish with a Yodel. I ate this combination after school for several years running. Why? Who the hell knows.

this is for you

Temptee Whipped Cream Cheese on Ritz Crackers, which is what Lil served. Don't follow the suggestion on the back of the box to

> " They say pimentos have something to do with olives but I don't buy that. "

put a pimento on top of the cream cheese. They say pimentos have something to do with olives but I don't buy that.

the story of R

Enormous shrink-wrapped corn muffin from cruddy deli at the corner of the street your therapist's office is on. Eat all eight trillion calories of it. You deserve it.

waiting

Leftovers off of the plates of people you don't know. Don't think too hard about it, just eat. Eschew anything that must be gnawed off a bone.

world peace

A tuna fish sandwich. Let the Gateses save the dolphins.

from the ashes like the phoenix

This is the essay to diet to. I was so depressed when I wrote it I couldn't eat. If you're a purger, go ahead.

jack has a thermos

If you can believe it, Jack sometimes puts ice and water in a thermos and brings it to the table at mealtime, which is insane. Anyway, that's all you get. Feel free to chew your ice, although ice chewing indicates an iron deficiency. Take iron pill.

is that what you're wearing?

By now you're pretty hungry. Have anything you want. Anything. As long as you put Hellman's Mayonnaise on it or dip it into a ramekin of melted butter. And when you are done, don't clean up.

they weren't brave

Eat your favorite food. Eat like this might be your last meal because you *never know*. What do you want? I'll tell you what *I* want. I want a box of ▶

Reading and Eating *(continued)*

Cap'n Crunch with a half gallon of skim milk, a Duncan Hines Deluxe II Butter Cake with Duncan Hines Classic Chocolate Frosting, and a pint of Friendly's chocolate chip ice cream, circa 1979, which also works well in front of the TV.

what happened after the chicken crossed the road

Put the book in your backpack or purse or whatever and go to Nathan's and have a hot dog and a paper dish of those greasy crinkly fries, which is what we always did after we ate dinner at my grandparents' house.

better safer warmer

Oy vey. Go to the Mamaroneck Diner on Mamaroneck Road in Mamaroneck, New York, or somewhere like it—although I hope there isn't—and have a blintz in honor of someone you loved.

hey!

Hey yourself. If you're pregnant, eat a box of saltines. If you're not pregnant, eat a box of saltines but spread cream cheese and strawberry preserves on them. If you're a man, go fuck yourself.

at the end of the day

Pack some nice snacks in a bag and take them to a person whose snacks might not be as nice as yours. Read on the bus on the way and eat some of the snacks because it's the thought that counts.

mountain men

Gorp.

megrim

Hard aged cheese, chocolate, and a Percoset, washed down with plenty of red wine.

passing

A bagel and cream cheese with smoked salmon. Or three or four gin and tonics and some crudités. Your choice.

gone home

Hard candies, squashed Junior Mints, packets of Equal.

the good swimmer or how I lost him

Pack a lunch box with a small carton of milk, an American cheese sandwich, a bag of Fritos, a Ring Ding, and some token carrot sticks. Eat alone.

horse kills owner or how I got him back

How can you eat while my child lies wheezing in the hospital? What are you, made of stone?

the few, the proud

Don't, under any circumstances, eat a truffle. But if you want a BLT or something, be my guest. ◡

Leave the Building Quickly
An Excerpt

Hi again. Just so you know, I have another book out now. It's called Leave the Building Quickly *and it's chockfull of brand new, all true stories. Yes, it is in hardcover, which is more expensive, I know, but you can't put a price on the appearance of being a book lover, which you only get from carrying around a hardcover.*

Donner Is Dead

A DEER HIT US. We were driving along, minding our own business when a deer jumped out of the woods, or maybe it jumped out of another car, who knows, and ran smack into us. We were on Route 7 driving north, about five minutes out of Bennington, on our way to celebrate "A Jew's Christmas in Vermont."

Christmas is just lovely in Vermont. Really, all the lights and the wreaths and the snow tipped steeples. You can light candles every night for a year and shake groggers until Haman rises from the dead, but Hanukah isn't fooling anyone. It's a diversionary tactic, kind of a "hey, over here" that Jewish parents employ to keep their children from feeling like those toys that get shipped out to the Island of Misfit Toys in *Rudolph the Red-Nosed Reindeer*. Everyone claims that Hanukah isn't meant to compete with the Christmas holiday. "It's a festival, it's a festival!!!" No one even knows when the hell it is. It changes every year. I'm not sure the *Farmer's Almanac* could predict it correctly.

> **We were on Route 7 driving north, about five minutes out of Bennington, on our way to celebrate 'A Jew's Christmas in Vermont.'**

Anyway, we didn't see it coming. All of a sudden, there was a very loud *thunk*, and then what seemed like a two-hundred-pound snowball exploded up and over the front windshield. My husband yelled, "What the fuck?" and I yelled "Fuck!"

We pulled over to the shoulder and sat for a moment in silence, trying to make sense of what just happened. It was a clear, bright night and very, very cold, maybe ten degrees. The air was thin; you could see for half a mile. There'd been nothing in the road. "I hope that was a deer," said David. "What else could it have been?" I asked. "I don't know," he said, "a person?"

David called the Bennington police and then, with some difficulty because the door only seemed able to open about ten inches, he got out of the car and walked around the front. Through the windshield, which, thankfully, was still there, I saw him mouth "Holy shit." He squeezed back into the car and said that the driver's side headlight was gone and the whole front left of the car was smashed in. The snowball effect we experienced must have been the glass from the headlight shooting up like sparks in the dark. The driver's side mirror was gone. David thought he saw some deer fur stuck to the ragged metal. Thank God.

There were police lights up ahead. A patrol car passed us by about two hundred yards, crossed the median, and came to a halt. The police were going to check on the deer first.

Here's what I don't understand: Why didn't Darwinism work for the deer? Cars have been around for what, over one hundred years, right? Why haven't the stupidest deer died out? Why isn't the gene that tells a deer to cross a four-lane highway obsolete? Why aren't ▶

> " Here's what I don't understand: Why didn't Darwinism work for the deer? "

Leave the Building Quickly (continued)

the smart deer at home in their beds making more smart deer?

What's wrong with the deer, I ask you? Zebras have stripes, for God's sake, giraffes have long necks. The leopards that survive in snowy climes are *white*. Was this an accident? No! The white leopards outlived the orangey ones because they were harder to see! And they made more white leopards and now, now, we have something called, yes, the *Snow Leopard*.

Where is the evolved deer? Haven't we waited long enough? They've certainly turned *tick-carrying* into a cottage industry. *That* didn't take long. Where is the deer that has a natural aversion to headlights? The deer who doesn't like the clickety-clacking noise its hoofs make on the asphalt? Where is the deer that doesn't like the way it feels to lay dying in the middle of the road, wondering what the fuck just happened? Where is he? You know where? In the middle of the road wondering what the fuck just happened, that's where. And while we're asking, what's on the other side that's so important to see at eleven o'clock at night? Better woods? ❧

The rest of this story and fifteen more stories from the "hilarious yet soul-touching Cynthia Kaplan" (my own quote) can be found in Leave the Building Quickly, *available in admirable hardcover from William Morrow, an imprint of HarperCollins Publishers, at your local bookseller in Spring 2007.*

D on't miss the next book by your favorite author. Sign up now for AuthorTracker by visiting www.AuthorTracker.com.